CONTROLLING PESTS AND DISEASES THE ORGANIC WAY

GARDEN MATTERS

CONTROLLING PESTS AND DISEASES THE ORGANIC WAY

JANE COURTIER

WARD LOCK

First published in Great Britain in 1992
by Ward Lock Limited, Villiers House,
41/47 Strand, London WC2N 5JE, England

A Cassell Imprint

Line drawings by Mike Shoebridge

Text filmset in 11/11½ point ITC Garamond Light
by Columns of Reading
Printed and bound in Great Britain
by HarperCollins Book Manufacturing, Glasgow

British Library Cataloguing in Publication Data

Courtier, Jane
Controlling pests and diseases the organic way. – (Garden matters)
I. Title II. Series
635.04996

ISBN 0–7063–7030–9

CONTENTS

CHAPTER 1

WHAT IS ORGANIC GARDENING?

Here we are, at the start of the book, and already we've got a problem. Just what do we mean by 'organic'? A dictionary definition is 'containing carbon' – all right for a chemist, but not much help to a gardener. Ask the man and woman in the street what they understand by organic growing and the phrases 'to grow without chemicals' 'natural' and 'non-poisonous' are likely to crop up, but most will find it difficult to be very specific. It's not surprising, as even ardent members of the organic movement can't agree on a definition!

Organic gardening is a very wide concept, and the nearest I can get to a sensible definition is 'growing in tune with Nature'. That means using naturally occurring, biodegradable products to feed plants and to protect them from pests, diseases and weeds. It means using pest control methods that are specific to the pest I am trying to get rid of, and which won't indiscriminately harm other creatures. It means treating my garden as an environment which is used not just for my pleasure, but as a home and means of existence for many other creatures.

At the same time, I want my garden to look attractive; to be a pleasant place to work and relax in, and to provide me with fruit and vegetables for the kitchen – so I need to exercise some control over what goes on there. But is it possible to do this without causing widespread disruption to the garden's other inhabitants?

Of course, you can go into any garden centre and find a vast range of chemicals that will sort out your problems at the squirt of an aerosol. So why bother looking for an organic alternative?

There are many reasons. The pesticides available today are very effective, but many people believe that they could also be harmful to humans and other animals. Manufacturers would dispute that, claiming that as long as the chemicals are used correctly, no harm will result. But what happens when the chemicals are not used according to the instructions, either through carelessness or an honest mistake? Even if pesticides were harmless to mammals, there is no doubt that few insecticides are selective. They may kill the aphids on your roses, but, with a few exceptions, they will also kill allies like bees, lacewings and ladybirds.

SYNTHETIC PESTICIDES

When synthetic pesticides were first used on a large scale, it was found that they often caused more problems than they solved. They killed the target insects very successfully, but they also killed predators that naturally controlled other pest species – and suddenly those other pests, without their natural checks, became seriously damaging for the first time. The balance of nature is a pretty effective method of pest control: after all, crops had been produced for thousands of years before synthetic pesticides were discovered. Once the balance of nature has been upset, though, it is very difficult to restore it.

Some chemicals are very persistent, remaining in the soil or plant tissues long after the original pest has gone, and residues of pesticides are found in nearly all the produce routinely sampled at supermarkets and greengrocers. These residues may be tiny, but do you **want** to eat pesticides, no matter how small the amount?

Insects (which account for most of our common plant pests) breed quickly and as a consequence they are able to build up resistance to certain poisons within a relatively short time. This means higher and higher doses are needed to achieve the same results, and eventually the poison may lose its effectiveness altogether. The more insects are exposed to the chemical – particularly low doses of it – the more quickly they build up resistance.

Pesticides do not always have a direct physical effect on non-target creatures, but may harm them in other ways. Bird populations may be reduced simply because widespread use of pesticide means there are insufficient insects for them to feed on. Pesticides enter the food chain: insects are sprayed, then eaten by small birds, which are eaten by foxes, for example. Each link of the chain receives a dose of the chemical the insect was sprayed with.

It is not only birds and animals that are affected by pesticides in the food chain, of course – man is, too. There is no doubt that when DDT was introduced, it was seen as a wonder chemical. It relieved poorer countries of the threat of famine, and the scourge of insect-spread diseases like malaria. However, it was not at first realized how persistent the chemical was; it is very efficiently stored in the body fat of animals that come into contact with it. Because it has been so widely used, virtually every man, woman and child in the world is believed to carry DDT in their bodies, having come into contact with it from pesticide residues on food crops, from consuming animals that have been contaminated within the food chain, or by other methods that are not yet fully understood. It has even been said that breast milk may be too contaminated with DDT to be the best food for babies!

The potency of modern pesticides can easily be seen by the side effects of lawn weedkillers based on plant hormones. These are selective weedkillers – they will kill broad-leaved weeds but, at the correct dose, not the narrow-leaved grasses that make up a lawn. Since most other garden plants are broad-leaved, great care must be taken not to allow any spray to reach them: tiny quantities can cause damage and death. Tomatoes are particularly susceptible. A man sprays his lawn with weedkiller; on the light breeze, minute particles of the spray drift into a neighbour's garden and through the open vents of his greenhouse. Soon the tomatoes exhibit typical symptoms of damage – malformed, fern-like foliage and characteristically pointed fruits. Such symptoms can be caused if the greenhouse owner stores his own weedkiller concentrate in the greenhouse: sufficient fumes may vaporize from even a tightly capped bottle to cause damage. And if, like a good organic gardener, you dig rotted farmyard manure into the border before planting your tomatoes, beware the straw in the manure that will often contain sufficient residue of hormone weedkiller to ruin your crop.

Waterways do not escape the chemical barrage, either. Pesticides applied to plants and soil are washed through the gound by rain, eventually finding their way to lakes and rivers where they may kill fish and other water-living creatures. Garden ponds have frequently been contaminated by 'safe' chemicals like derris – safe to mammals but fatal to fish. Fertilizers, applied in heavy doses to the soil and fields, seep into waterways and encourage such luxuriant growth of algae that it robs the water of oxygen and suffocates the animals living in it.

If you have young children, you may be particularly concerned about using chemicals: any harmful effects are multiplied in young, rapidly growing bodies. A toddler is not capable of understanding the 'harvest interval' that must be observed between spraying and picking certain crops, so you must make sure he or she does not sample your recently sprayed fruit. And unless you are extremely careful, there is always the chance of an accident happening as curious children explore bottles and packages which have not been as securely stored as you thought: much-loved pets, too, have come to grief from the contents of the garden shed.

We have learnt from our past experiences, of course, and new pesticide regulations mean that today's chemicals are far more rigorously tested for safety before they can be marketed. They can only be tested for the possible side effects we are already aware of, however: who knows what might be discovered in the future?

NATURAL PESTICIDES

Most people feel that a naturally occurring substance is far less likely to have damaging repercussions for the environment than a synthetic chemical, though this does not necessarily mean that it is harmless. Nicotine is approved by organic gardening associations as a pesticide because it is derived from plant material – but it is very toxic to mammals, too. I'm not sure that many gardeners would feel happy about using a brew concocted from soaked cigarette butts to spray their plants, as is sometimes recommended (see page 93); in some countries this sort of home-brew is now illegal anyway.

I have no doubt some dedicated and knowledgeable organic gardeners will find plenty to argue with in this book. But what I aim to do is suggest how to keep your garden looking good and cropping well in ways that are safe – safe for us **and** the environment. You will have the benefit of a garden full of interest, from birds, butterflies, hedgehogs and all sorts of other creatures, as well as the plants you grow. You will be able to harvest your vegetables and fruit in the knowledge that they are not full of some chemical which may turn out to have an adverse effect on your health.

Hopefully, you will also find you develop a better understanding of the way plants and animals grow and interact, and as a result gardening will become more enjoyable and more fulfilling. Growing without chemicals does not necessarily mean a lot more work, either; in fact, you may find it a lot easier since you will be relying on Nature to do much of the work for you.

In return, you may have to accept a garden that is perhaps a little short of perfect. A few holes in the cabbages, the odd greenfly on the roses – the natural garden relies on keeping the damage down to acceptable levels rather than eliminating it altogether. But if you're new to organic gardening, you may be surprised at just how effective and satisfying it can be.

CHAPTER 2

HELP AT HAND

Fortunately, there are plenty of creatures in the garden who will help a great deal in the battle against pests. Note, however that it's not always easy to tell who is friend and who is foe – the same creature can be both at different times of the year!

A pest is a creature that causes significant damage to the plants we are cultivating. A beneficial creature is one that helps to limit that damage by natural means, usually by preying on the pest species. By beneficial creatures, I include only those that occur naturally in the garden, not those non-indigenous species that are deliberately introduced by the gardener: these are called biological controls, and I'll be looking at them separately (see Chapter 5).

A number of beneficial creatures are insects; others are mammals and birds.

BENEFICIAL INSECTS

LADYBIRDS

These small beetles must be the best known of the garden helpers. The red, black-spotted species – two-spot or seven-spot – are very familiar, but there are also other types, including black with red spots, black with yellow spots and yellow with black spots.

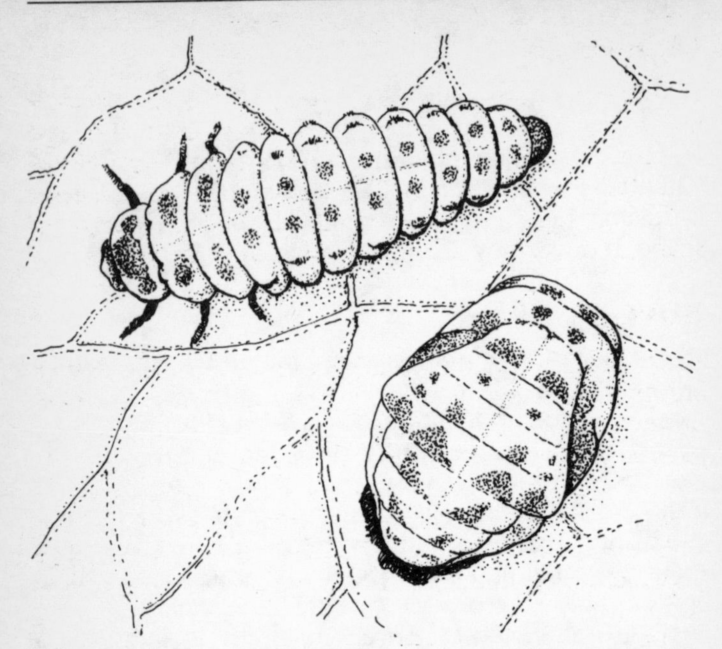

Fig. 1 *Ladybird larvae look like miniature armadillos, while their pupae are sometimes mistaken for Colorado beetles.*

Perhaps not so easily recognized is the larval stage of the ladybird (Fig. 1). These small, slate-blue creatures look a little like tiny armadillos, with four orange spots on their sides. Both adults and larvae eat large numbers of aphids: each larva consumes up to 500 in the three weeks before it turns into an adult. Aphids are their most popular food, but other soft-bodied insects are eaten as well. The yellow pupa of the ladybird is sometimes mistaken for a colorado beetle.

Ladybirds have been accused of giving an occasional nip in irritation when stranded on human flesh, but they are entirely harmless. In years when aphid infestations are bad, vast numbers of ladybirds are also produced and, when allowed, will deal with aphids very efficiently.

LACEWINGS

Delicate creatures with pale green bodies, large, lace-like wings and long, questing antennae, lacewings are also responsible for the destruction of aphids. Both larvae and adults feed on them. The larvae are inconspicuous on plants, but look rather similar to ladybird larvae in shape, though they are pale brown in colour.

HOVERFLIES

There are several different species of hoverfly, which are often mistaken for bees or wasps. This gives them protection from birds and other predators, who are not keen to risk getting stung, even though the hoverflies do not, in fact, have stings. As their name suggests, these flies can hover virtually stationary in the air, unlike bees and wasps: they are also easy to distinguish when at rest because, being flies, they have only one pair of wings and not two.

The small, worm-like larvae of most hoverflies feed on aphids, though the larvae of the narcissus fly, which looks a little like a small bumble bee, feeds on daffodil bulbs and is definitely not as welcome in the garden as the rest of the family.

WASPS

The common wasp, while being a well-known pest to picnickers, fruit growers and bee-keepers at the end of the summer, is, for the rest of the year, a useful ally in the garden. All the time young are being reared, wasps search endlessly for small, soft-bodied insects like caterpillars and aphids as protein sources for the grubs. Only when the young wasps have matured do the adults spend their remaining days gorging on sweet substances (and becoming more likely to sting humans into the bargain).

The main friends of the gardener, however, are the parasitic wasps. There are several species, generally unnoticed in the garden. They tend to have long, slender bodies with a pronounced waist, and long legs and antennae. They do not prey on insects directly, but lay eggs into the bodies of aphids, caterpillars and other insects. When the eggs hatch, they feed on the host from within. Sometimes only one egg is laid in each host, sometimes dozens. Ichneumon, braconid and chalcid wasps are some of the commonest parasitic wasps.

BEETLES

The majority of the huge family of beetles are helpful, or at least harmless, in the garden, though there are a few that are notable pests. Among the helpful ones are the ladybirds we have already looked at, and ground beetles; some 350 species of them. As their name suggests, ground beetles live on the soil surface, especially around organic matter and plant debris. Although they have wings they tend not to fly, but they may climb plants and trees in search of prey. They are mainly large, black or dark brown, and do much of their hunting at night. Both larvae and adults eat insects (and their eggs), slugs and earthworms.

Useful beetles are the rove beetles, best known of which is the devil's coach horse (Fig. 2). This is a rather fearsome-looking beast because of its habit of arching its tail like a scorpion when threatened: it is quite harmless to humans, however. Like the ground beetles, rove beetle adults and larvae feed on soil insects, eggs and grubs.

CENTIPEDES

These useful pest predators suffer because many people confuse them with millipedes, which are damaging to plant roots. Centipedes tend to be golden brown in

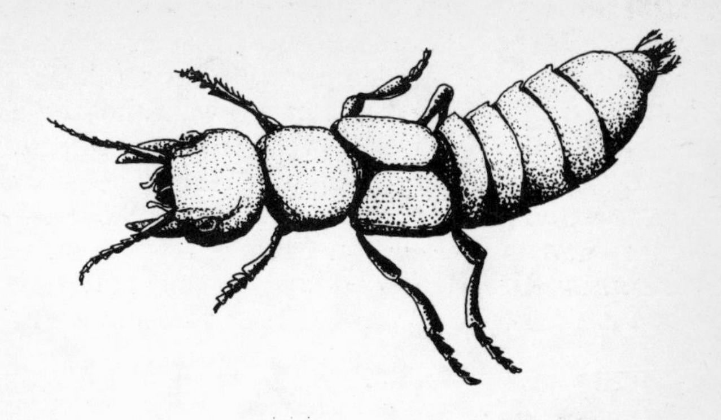

Fig. 2 *The devil's coach horse may look fearsome, but it is harmless to humans.*

colour and run about very quickly; millipedes are usually black or very dark brown (though some are almost the same colour as centipedes) and roll up rather than run away when disturbed. Centipedes have fairly long legs, one pair to each segment, which can be seen moving individually; millipedes have shorter legs and many more of them – two pairs per segment – and they move like a wave down the creature's side.

Centipedes feed on a variety of insects, insect eggs, small slugs and earthworms, and will chase after their prey at speed.

EARWIGS

One of those creatures, like the common wasp, which leads a double life. Most gardeners consider earwigs a nuisance as they feed on dahlia and chrysanthemum blooms, often distorting and spoiling the flowers. However, they feed on insects (including aphids), as well as on plant material, and if you don't grow dahlias or chrysanthemums you are unlikely to find earwigs a problem.

CAPSIDS

Capsid bugs are best known as pests, but the black-kneed capsid, among others, is a predator that will feed on red spider mites, leaf hoppers, aphids and other pests of fruit trees. Indiscriminate presticide use will wipe out this helpful ally along with the pests.

ANTHOCORID BUGS

These reddish-coloured bugs look like small capsids and prey on a range of pests, particularly of fruit trees. Like capsids, they are badly affected by the use of tar oil and similar winter washes applied as preventive treatment against pests.

BEES

Most gardeners appreciate that bees are useful in the garden. They do not have a direct effect on pests: while the grubs of wasps feed on insects, bees rely purely on pollen as a protein source for their young. However, they are very important as pollinators, without which there would be little in the way of fruit crops (or of fruiting vegetables such as marrows and beans). Avoiding the use of insecticides, or timing their application very carefully, is necessary to avoid killing bees and the other many pollinating insects.

OTHER HELPERS

HEDGEHOGS

Most people now know that hedgehogs are of great benefit in the garden, as slugs are their staple diet. They are nocturnal, and can often be heard, rather than seen, shuffling about the garden in search of food. Don't give them bread and milk to encourage them – it upsets their digestive system and can be fatal. Tinned cat food is a better bet.

BIRDS

Jekyll and Hyde characters again: birds can be a nuisance at times, but their good side far outweighs their bad. Not all birds eat pests, but insects form a major part of the diet of several species, notably the tit family. These small birds will painstakingly pick aphids from stems and in the crevices of tree bark: they are also extremely partial to small caterpillars, especially when feeding their young. Robins are well known for hopping along behind whenever there is any winter digging going on: their sharp eyes soon spot caterpillars and other soil pests turned up by the spade. Thrushes will help keep down those troublesome pests, snails and slugs; and starlings will eat almost anything, including large numbers of insect pests. Many birds also eat worms, which is not so helpful, and berries and fruit may need protecting from them in season.

HENS AND DUCKS

Hens particularly may ruin your vegetable crops if allowed to run riot, but hens and ducks can be very useful for controlling pests. Ducks are partial to slugs and caterpillars, and will pick caterpillars off cabbage leaves with great concentration. Hens will scratch up bare soil to expose and exterminate soil-living pests, and pick aphids and other insects off plants.

SPIDERS

Spiders are, like several natural controls, indiscriminate

about their prey, and catch beneficial insects as well as damaging ones. They do dispose of quite a number of pest species, however. Once the prey is trapped, it is paralyzed by the spider's poison fangs and often stored for eating later.

Web-forming spiders are best known, but there are species that don't form webs at all. Wolf spiders lurk amongst leaf litter and low-growing plants at soil level, ready to give chase to any likely looking snack that happens to amble by.

Harvestmen are not spiders but belong to the spider family: like them they have eight legs instead of an insect's six. Harvestmen have small bodies consisting of just one section, unlike the spider's two, and very long legs on which they sway when disturbed. They prey on a number of small, soft-bodied insects.

Mites, very small creatures also related to spiders, are notorious because of a few wayward members of the family that cause a great deal of damage to plants. Other mites, however, are predatory and are definitely on the gardener's side. Fruit trees are sometimes attacked by the pest 'fruit tree red spider mite', but several species of predatory mite also occur wherever it is found. Precautionary winter washing of fruit trees with insecticide cuts down the number of overwintering pest mites, but also kills the helpful mites that may well have kept the situation under control if allowed to do so.

TOADS AND FROGS

A wide range of insects, plus slugs, snails and worms (and even the occasional small mammal) make up the diet of these amphibious creatures. They may visit your garden even if you don't have a pond, as water is only necessary for them at certain times, but a pool will certainly encourage them into your garden in greater numbers.

Toads are dry, rough-skinned creatures, a dull brown in colour. When they jump, they hop using all four legs as springboards. Frogs have a smooth, shiny, moist skin which is a yellowish green. Their back legs are very long and used to propel themselves forward more gracefully and efficiently than toads.

Both frogs and toads need moist, shady places in which to shelter, and a pool of some sort – it needn't be big – in which to breed.

BATS

These fascinating creatures are declining in Britain, mainly due to the treatment of roof timbers against woodworm: bats are very sensitive to the pesticides used. They eat large numbers of moths and other night-flying insects: they can be seen at dusk silently flying backwards and forwards overhead wherever insects congregate. Some types of bat also fly during daylight hours. They locate their prey using radar and are remarkably accurate.

The commonest bat is the pipistrelle, with its attractive, mouse-like face; long-eared bats are rarer, and several other species are now very rare indeed. They have separate winter and summer roosts, usually under the tiles of old sheds and houses. They are protected by law, so if you are lucky enough to have them roosting on your property, you must not interfere with them in any way.

CATS

Another creature that is more often than not considered a pest, as it scratches up seedbeds and claws at tree trunks, but cats do have some uses in the garden. They catch a variety of rodents that would otherwise damage your plants, and are often successful at catching moles – an animal which it is very difficult to control successfully by any means, organic or otherwise. If you have your own cat, he will defend his territory from other cats in the area and you may suffer less feline depradation than catless gardens.

ENCOURAGING NATURAL HELPERS

While many of these beneficial creatures are present in most gardens, you can get more than your fair share of them by taking steps to make your garden particularly attractive.

THE POND

This is one of the best features you can provide. It will attract frogs, toads and newts, birds (to drink), and a variety of useful and attractive insect life.

Ponds do not have to be very large, though the larger they are, the easier they tend to be to keep looking nice. Choose an open site which is not overshadowed by trees, particularly deciduous trees which will foul the the water with leaves in autumn. Make the pool at least 45 cm (18 ins) deep, to avoid it freezing solid in hard weather, and make some shelves round the edge for marginal plants.

The best lining for the pond is butyl rubber: it is not as cheap as polythene but is very long lasting (it has a better life expectancy than concrete). Plant oxygenators such as *Elodea canadensis* in the bottom of the pond to keep the water in good condition; you can obtain these from most large garden centres. There are lots of other water plants you can add: choose one with floating leaves, like a water lily, to shade the surface; another with spiky leaves, like a rush, for dragonfly and other insect nymphs to crawl up; and a range of marginal plants to plant up a transitional boggy area between the pond and dry land.

Make sure there is an escape route for animals which may inadvertently fall into the pond – hedgehogs are adept at doing this! Have one gently sloping side where they can scramble out, or provide a piece of wood as a bridge to dry land.

The true wildlife pond should not have fish, which feed on insect larvae and other wildlife – but on the other hand, they do control midges, gnats and mosquitoes, which might not be so welcome in the garden on a summer evening.

When the pond is first filled, the water very quickly turns bright green, or sometimes red, as an explosion of algae exploit the new habitat. Do not try to remove or control this algae; it is harmless and provides a food source for some pond dwellers. As the water matures, so the algae will settle down, though a small, shallow pond may never have perfectly clear water. If a 'pea soup' pool worries you, make your pond at least 2 m (6½ ft) across; then you are more likely to achieve a good natural balance, with clear water.

To establish frogs in your pond, import some frogspawn from a neighbour's pond in spring, or contact a local wildlife group, who can usually put you in touch with someone to help. Don't take frogspawn from wild ponds.

SECLUDED SHELTER

A pile of old logs provides home and shelter for a number of animals including various rodents, grass snakes, and all sorts of insects, which in turn attract insect-eating birds. Heaps of stones, particularly large, flattish stones, will also encourage grass snakes, frogs and toads, as they maintain a cool, moist habitat. And you have only to turn over a large stone in the garden to see what a variety of insects scuttles away! If you are lucky enough to have a resident toad in your greenhouse, provide him with a large clay pot turned on its side and give it an occasional watering; allow a few weeds to grow round it for preference.

Leave your log and stone pile undisturbed, tucked away in an unobtrusive corner of the garden if you think it looks untidy. Don't forget that the compost heap also provides an attractive dwelling for many creatures – another good reason for making one. Heaps of leaf litter, dry straw and bracken, and twiggy branches are also useful.

Unfortunately several less desirable creatures will find your wildlife habitats appealing, too, so you are bound to end up giving shelter to some pests. However, the good will generally outweigh the bad.

WINTER HOMES

Climbing plants which form a thick mass of old growths on walls and fences provide perfect winter shelter for many small mammals, insects and birds. Don't be in too much of a hurry to 'tidy up' climbers once summer is over. Plant evergreen varieties that are attractive all year round: variegated ivy is an ideal choice.

In beds and borders, too, leave the dead growth on herbaceous plants instead of cutting it all down to soil level in the autumn.

THE RIGHT PLANTS

The way in which you plant your garden up has a great effect on the amount of wildlife visiting. As long as space allows, choose at least one native tree; trees provide cover, can support vast numbers of insects and encourage birds to the garden.

Willows are often recommended as good trees for wildlife, but they are not good for gardens. Their roots spread too widely and can clog drains and cause structural damage to houses, especially on clay soil. The silver birch is a good alternative; it is quick growing, not fussy about soil, attractive to look at and supports a good range of insects without being a host to too many pests. For an evergreen, holly is a good choice, or a native conifer such as Scots pine.

Plants that bear cherries will be welcomed by birds and rodents in the winter. Pyracantha and cotoneaster are common, and attractive to both wildlife and the gardener. Crab apples will also provide a colourful show of fruit which will eventually be consumed by birds.

Flowering plants are visited by insects for their nectar and pollen. Choose single flowers rather than double varieties, which often produce no nectar at all. Try to stretch out the flowering season: willow, blackthorn, crocuses and so on are useful because they flower very early in the year and will provide nectar and pollen when it is otherwise scarce: ivy, on the other hand, is one of the latest, and is busy with the hum of bees and hoverflies on sunny, late autumn days.

Buckwheat is excellent for attracting hoverflies, as is *Limnanthes douglasii*, the poached egg flower; sedum, buddleia and Michaelmas daisies are alive with butterflies in the summer. Don't forget food plants for butterfly larvae – the common stinging nettle is a great favourite, and such a patch could always be left somewhere in the garden.

FEEDING TIME

One of the most obvious ways to encourage wildlife is by supplying food, particularly for birds (Fig. 3). Feed them only in the winter and early spring: for the rest of the year they can find their own food. Water should be provided all year, however.

Site a bird table where cats cannot creep up on

Food plants for butterflies

Brimstone: buckthorn
Brown hairstreak: blackthorn
Clouded yellow: clover and lucerne
Comma: hop, nettle, currant
Common blue: bird's foot trefoil, rest harrow
Fritillaries: dog violet
Large skipper: cocksfoot grass
Orange tip: garlic mustard, lady's smock
Painted lady: thistle
Peacock: nettle
Purple emperor: sallow
Purple hairstreak: oak
Red admiral: nettle
Small copper: dock
Small tortoishell: nettle
White admiral: honeysuckle

feeding birds unnoticed; the centre of a lawn is usually the best spot. A tree fairly near by will give the birds some cover to fly to.

Provide a range of food for different birds. Wild bird food purchased from shops will suit seed eaters: scraps of cooked vegetables, soaked bread, stale cake, chopped meat (not too much or you may attract rats), suet, bacon rinds, grated stale cheese, cooked rice, oats, stale breakfast cereals, dried fruit and damaged stored fruit such as apples should keep everyone happy. Peanuts in special containers will be eaten by tits, greenfinches and nuthatches; and as an occasional treat, mealworms (from an angling shop) will please the robins. Fresh coconut hung up in the shell is good, but don't put out dessicated coconut.

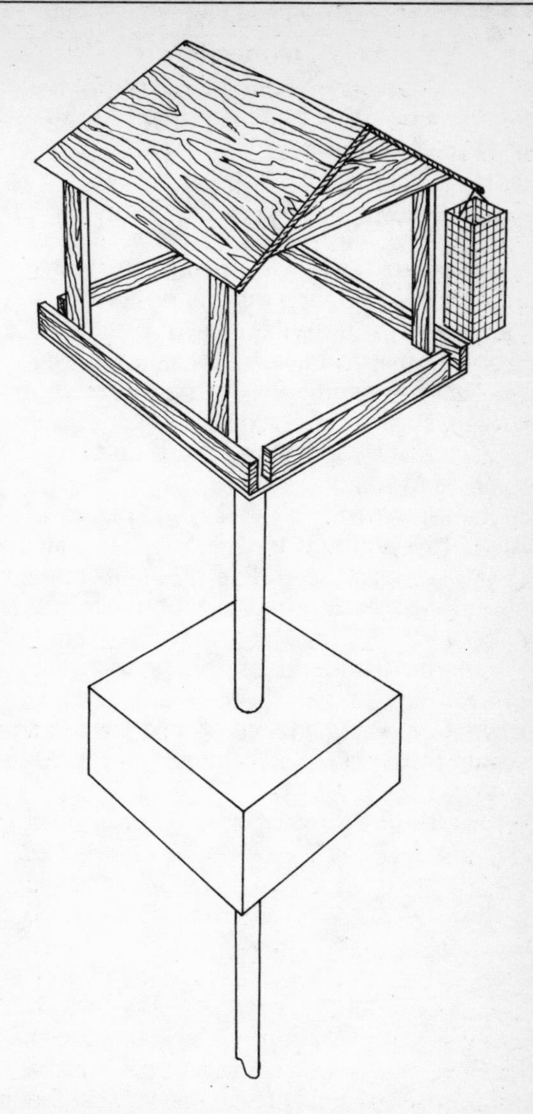

Fig. 3 *Encourage birds into the garden with a bird table. Feeding is unnecessary during the summer.*

Once you have started feeding, continue to do so as birds will come to depend on it. Put out small amounts that are cleared up within a couple of hours rather than lying around for days.

PURPOSE-BUILT HOMES

Nest boxes for birds will encourage them to breed in your garden. There are various designs to suit different species: apart from the familiar tit boxes, there are ones for robins, flycatchers and wagtails; swifts; swallows; house martins and owls. They are quite easy to construct, and several wildlife organizations can give you instructions for making these and bat boxes. They also sell ready-made boxes.

Hedgehogs need a sheltered place to hibernate between late autumn and early spring, and will often choose a compost heap, leaf pile or – dangerously – a pre-lit bonfire stack. Provide a home in the shape of a box about 40 cm (16 in) square and 30 cm (12 in) high, buried under a mound of soil or leaf litter. It should have an access tunnel about 60 cm (2 ft) long, formed of bricks supporting a piece of wood as a roof. Put some dry straw in the box as bedding.

CHAPTER 3

CULTURAL TECHNIQUES

The organic control of pests and disease does not just mean finding a harmless way of exterminating them; it also means trying to prevent them occurring in the first place. There are lots of things you can do to protect your plants from attack.

SOIL FERTILITY

Healthy plants are not, as is often stated, less likely to be attacked by pests. They will, however, be better able to withstand pest and disease attack than a plant which is struggling to grow in unsuitable conditions.

It is very important to build up your soil fertility so that plants can get all the nutrients they need. They must also be able to get adequate supplies of water (without having their roots damaged by excess). A deficiency of nutrients, drought, or waterlogging will put plants under stress and render them less capable of fighting off pests and diseases.

Soil types vary greatly around the country, and from garden to garden within a small area. You must first decide what type of soil you have. You will probably already have a good idea from the way your plants grow at the moment: you will know whether they need copious watering in the summer to stay alive; whether the soil is so sticky and heavy you can hardly work it when it is time to sow seeds in spring; whether azaleas and rhododendrons flourish or die; whether plants grow tall and lush or remain small and pinched unless given fertilizers.

The two extreme types of soil are sand and clay, with a whole range in between. *Sandy soils* are very light, and water and plant foods drain through them very quickly. If you rub the soil between your fingers it will feel gritty, and if you try to roll a lump of damp soil into a ball, it will fall apart. *Clay soils* are sticky, and feel slippery when rubbed between the fingers. They can be rolled into a ball and shaped into a rope when rolled between the palms, without crumbling or breaking up. While sandy soils warm up quickly in spring and are very easy to work, clay soils remain cold and wet, and are difficult to break down to the small crumbs (called 'a fine tilth') we need for sowing and planting.

Most soils consist mainly of *loam*, fibrous, dark material, and either sand or clay (or sometimes both) in varying proportions. The ideal soil contains sufficient sand to allow surplus water to drain away and to make it easy to break down to fine crumbs fairly early in the spring: it should also contain enough clay to hold on to water in times of drought, and to keep nutrients within reach of the plants' roots instead of letting them be washed out of the soil by rain.

The other important ingredient of the ideal soil is *humus*. This comes from organic matter, and is a spongy material that is something of a miracle worker. It helps sandy soils become more nutrient- and water-retentive, and it helps clay soils drain more efficiently and be broken down more easily – so whichever type of soil you have, organic matter will improve it.

Garden compost, farmyard manure and leafmould are the three most important sources of organic matter; others include spent mushroom compost, spent hops, and seaweed, as well as the various proprietary 'bagged' manures. These usually consist of chicken or other animal manure which has been composted and partly dried to make it compact and easy to handle.

Peat is not a good source of humus for the organic gardener. Apart from the fact that it contains no nutrients, which many of the other sources such as compost do, it is a dwindling natural resource, and important wildlife habitats are being destroyed by peat digging.

GARDEN COMPOST

Nearly all soils are greatly improved by adding large quantities of organic matter, and one of the easiest and cheapest ways to provide organic matter is by making garden compost. Apart from providing a valuable product, it solves the problem of what to do with garden refuse: most of it can be composted, though not all.

A badly made compost heap can be a rather unattractive sight, but don't let other people's mistakes put you off. A properly made compost heap does not smell, or attract flies or vermin, and making compost properly is not at all difficult.

Why compost?

Before organic matter can do any good in the soil, it has to be broken down by bacteria. In order to do their job, the bacteria need a supply of nitrogen, and this will be taken from the soil if fresh organic matter is added to it. This could mean crops going hungry until the organic matter is broken down – and the breakdown will be much slower than on a compost heap. Well-rotted compost will add nutrients to the soil straight away, and is also much easier to transport and incorporate into the soil than fresh material.

Apart from anything else, fresh organic matter spread on the surface of the soil does not look at all attractive.

Essentials for composting

All organic material will eventually rot down. Good compost needs a supply of organic matter, the right bacteria, moisture, warmth and air. If there is insufficient air, a different type of bacteria (anaerobic bacteria) will start to work on the organic matter, and the result will be more like silage than compost – slimy, smelly and unpleasant.

First you need to construct some sort of container to keep the compost tidy; while a container is not essential to make compost, it does make the job easier, more effective and neater. There are many types of container to choose from.

Platic bag

You can put a good mixture of waste into an ordinary black plastic dustbin liner. Punch a few holes in the base for drainage, fill the bag and tie the top loosely. This method is cheap, but the compost does not always get enough air, and often becomes slimy and unpleasant. It is essential to put a good mixture of plant material in the bag, including some bulky items (such as straw) to provide air.

Plastic bin

There are various styles of these on the market. Most are like plastic dustbins but with no base – you lift them off when the compost is ready. This is a lot easier than trying to shovel the finished compost out of an ordinary dustbin! Some have holes in the sides to allow in extra air.

Their drawbacks are that they are not really big enough for the most efficient rotting, and can be quite expensive for what they are.

Mesh containers

Plastic-covered wire mesh or even just chicken wire can be used to keep the heap in place but allow too much heat to escape for efficient composting. The proprietary types are again often too small to be much good.

Rotating drum

A plastic compost bin that can be rotated to mix and aerate the contents. It is claimed to speed up rotting, but is quite expensive.

Wooden bin

These can be bought ready made, or constructed quite easily (and more cheaply). The wood should be treated with a wood preservative. Leave small gaps (about 1 cm) between the planks to allow air to the heap. Make the front of wooden slats which slide into place and can be removed so that you can get at the compost easily. Minimum size for the container is about 1 m in each direction – the length and width should be more if possible, but don't make the heap too high or you will find it difficult to turn.

A double bin is useful; once the first heap has reached the top of the bin it can be turned into the other side, then left to mature while the next heap is built up.

Wood can be fairly expensive but should last well and has good insulating properties to keep the heap warm.

Scrap materials

A perfectly good compost container can be made out of pieces of scrap wood, corrugated iron, rigid plastic, chipboard, plywood and so on. If you build the heap correctly there will be sufficient air within it – the container needs to provide insulation rather than aeration, so don't worry too much about leaving gaps for air supply.

Where to put the container

While composting is speeded up by warmth, the decomposing materials generate their own heat. It is sufficient to conserve this warmth – you don't need any extra. If a heap is made in a sunny position, it is likely to dry out excessively, so choose a shady spot. No-one likes to make their compost heap the centre of attraction; it is usually sited in an unobtrusive corner, or screened off by a hedge or wall.

At the same time, it should be accessible. Ensure that you can approach it easily with a full wheelbarrow, and that there is room to fork the debris off the barrow on to the heap and turn the barrow round again. Try not to make it miles away from the main garden, or lawn mowing could become a real chore when there are innumerable trips to make with the grass clippings. The heap should be built directly on the soil, not on slabs or concrete.

What to use

Most gardens provide plenty of organic material. **Plant debris** – dead leaves, spent crops, soft prunings and so on – can be used, but don't include diseased plants or parts of plants as the disease spores are likely to survive the composting process. **Lawn mowings** can be used with care: because they are densely packed, air cannot penetrate and the clippings become slimy and unpleasant. Always mix lawn mowings with other material (straw is good) to improve the aeration.

Woody material will not rot unless broken down into small pieces. Woody prunings can be chopped up with secateurs, but it is a time-consuming job: if you have a lot of woody material to dispose of you might find a shredder useful. These can be hand operated or can have powered blades: if you don't want to buy one they can be hired from tool hire shops.

Always use eye protection (safety glasses) with shredders as small pieces can be thrown up by the blades; use gloves when feeding material into the hopper. Be careful not to put in stones, soil or glass by accident.

Kitchen waste – fruit and vegetable peelings, tea bags and so on – can be used, but avoid cooked food and meat, which could attract vermin. **Newspapers** can be added in small quantities; either shredded or in single sheets between other material.

Weeds are fine, but do not include perennial weed roots or seeding weeds. While a compost heap may heat up enough to kill seeds, it is by no means certain that it will, and you could easily end up spreading weeds round the garden.

Spent compost from containers, growing bags etc. can be added to the heap; **spent hops** can often be obtained cheaply from breweries, and **seaweed** is a good additive for those who have no problem transporting it from beaches.

Animal manure is an excellent additive, speeding up the breakdown of the other materials and increasing the nutrient value of the compost. Unrotted manures can be too "hot" if used on the garden in their fresh state, scorching plant roots.

Making the heap

Start with a layer of coarse material such as strawy manure or bulky weeds so that there is good aeration at the base of the heap. Add more material – vegetable peelings, weeds, etc. – until a 15–23 cm (6–9 in) layer has been formed. You can then add a sprinkling of nitrogen fertiliser, proprietary compost activator (which is much the same thing), animal manure or a thin layer of garden soil. The nitrogen this provides helps to feed the bacteria which are starting to break down the material. Build up another 15–23 cm (6–9 in) layer, then dust it with garden lime. Continue like this until the container is full (remember that the compost will shrink as it rots, so you can top up the container over several weeks).

Turn the heap occasionally to mix the contents

thoroughly and allow air into it. Try to mix sides to middle and top to bottom; easier said than done, but there's no need to be too much of a perfectionist about it. If you have a double bin, when the first container is nearly full, fork it out into the second bin, turning it as you go. The more frequently compost is turned, the quicker it will rot: the optimum is every few days, but few people will have the time, enthusiasm or strength top do this. An occasional turning is better than nothing, but even a heap that is not turned at all will rot down eventually, though it may do so a little unevenly.

Usually fresh plant matter will supply enough moisture to satisfy the requirements of the bacteria, but in dry weather, or if you have used dry materials like straw, the heap may need watering. You should give just enough water to moisten it throughout: it should not be wet as this will slow down decomposition.

When the container is completely full, use a square of old carpet, a piece of heavy duty polythene, or a piece of hardboard or similar, to form a lid. This helps keep in heat and keep out excess rain.

Start another heap while the fist one matures. In the summer, compost may be well rotted in two months; in the autumn, it is likely to take until the following spring before it is useable. It should be dark and crumbly, with a pleasant 'earthy' smell to it, and not too much identifiable plant matter. If your heap hasn't rotted perfectly when you come to use it (often the middle is fine but there may be dubious bits round the edges) simply add the unrotted parts to your current heap and use the rest.

Rotted compost can be applied to the soil virtually any time; during the winter digging is usually most convenient.

ACID OR ALKALINE?

One more thing to know about your soil is whether it is

acid or alkaline. Alkaline soils contain lime (chalk), which has the effect of making certain nutrients, particularly iron, unavailable to plants, even though they might be present in the soil. Some plants are very sensitive to the presence of lime and will not grow satisfactorily in alkaline soils. Rhododendrons and azaleas are probably the best-known lime haters, but there are plenty of other choice shrubs that must be avoided if you garden on chalk.

You may be able to see lumps of chalk (often with flints) in your soil; you can also gain a clue as to soil acidity by noting the plants that grow naturally in the region. Beech woodlands mean you are likely to have an alkaline soil: rhododendron thickets and heathland indicate acidity.

A soil pH testing kit (Fig. 4), will help if you are not sure. Samples of soil from different areas of the garden

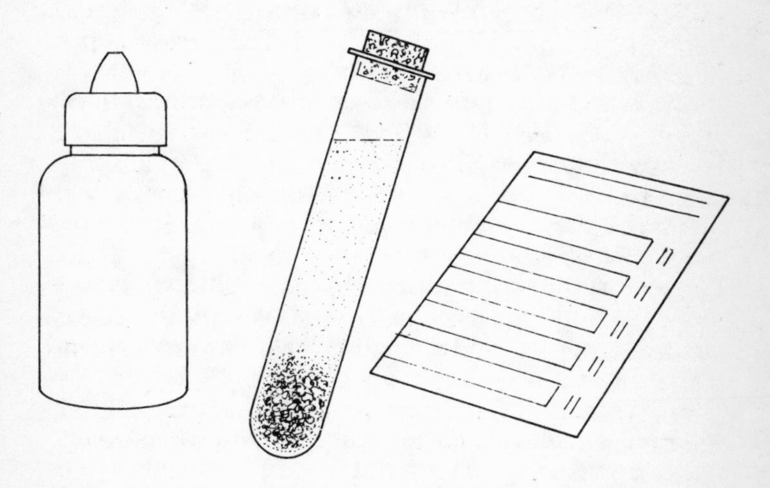

Fig. 4 *A simple testing kit will tell you the acidity (pH) level of your soil.*

are taken from just below the surface and shaken up in a test tube with a measured amount of a reagent. The soil is left to settle out, or the liquid is filtered off. The colour of the liquid is compared with a chart supplied with the kit, which will indicate the pH level of the soil. pH7 is neutral – neither acid nor alkaline. Anything higher than pH7 is alkaline, while anything lower is acid.

Acid soils can be made more alkaline by the addition of garden lime: alkaline soils are more difficult to alter. Flowers of sulphur has a temporary effect, but the method of working with alkaline soil is not to try to grow lime-hating plants (except in tubs or special beds of imported acid soil) and to give special feeds of trace elements and sequestered iron (which can be absorbed by plants despite alkaline conditions) where necessary.

RESISTANT VARIETIES

It makes sense to grow those plants and varieties of plants which are either less likely to be attacked by pests and diseases, or which are better able to withstand them if they are attacked.

You may know from your own experience some plants that always seem to do badly for you. These may not be the same for everyone – the locality and even the microclimate in your garden will give different results to those of other gardeners. First, choose plants that are suited to your soil and climate: trying to grow those that are not suited may be a challenge, but be prepared for disappointment. Make notes of how all new plants you try fare; it is difficult to remember in the winter, when you are making plans for next year. Where plants are known to be difficult, try adapting your growing techniques as suggested in this section (choosing the right sowing time to avoid pea moth, for instance) before giving up on them altogether, but if you have persistent failures it is probably best to admit defeat and find something else to grow.

The charts later in this book (see page 61) will help you identify potential problem plants. A great deal of work has been done in breeding varieties of plants that have greater resistance to certain diseases and, to a lesser extent, pests: read the new seed catalogues each autumn to find them.

KEEPING YOUR EYES OPEN

Being observant is especially important for the organic gardener. Take time to look at your plants closely and see how they are growing. Do they look healthy? Are they uniform, or are some plants in worse conditions than others, and if so, why? Look closely for the first signs of pests or diseases, making sure to check the undersides of leaves, which is where problems often start. An attack can often be literally nipped in the bud by the observant gardener; control measures are always more successful if they are carried out before too much damage has been done.

SIMPLE CONTROL TECHNIQUES

HAND PICKING

This can be surprisingly effective in the early stages and prevent the need for more complicated or even damaging measures.

Caterpillars, particularly cabbage white caterpillars, are big enough to be picked off plants and crushed (or fed to the chickens). Where leaves are more heavily infested, pick off the whole leaf but don't strip the plant. Aphids usually infest the soft young growing tips of plants: these can be pinched out and destroyed. If this would spoil the shape of an ornamental plant, the infestation can often be washed off gently in a bowl of warm water. If you find one heavily infested shoot tip, the rest are often likely to be infested shortly afterwards, so be on the alert.

WINTER DIGGING

This will expose soil-living pests to birds: that's what that friendly robin is hoping as he hops along behind you. *Tidying away* debris and weeds will cut down hiding places for some pests, but will also decrease the number of natural predators in the garden, so needs to be carried out carefully. Keep the area around particularly susceptible plants clean, but leave a bit of untidiness where it doesn't matter too much.

ALTERNATIVE HOSTS

Providing alternative hosts for pests can make sure they stay around your garden. Various pests retreat from crop plants to a different plant on which to spend the winter – for instance an aphid that infests carrots spends the rest of its time on willows: bean aphids retire to the spindle tree (euonymus). If you have a particular problem with these pests, ensure that you are not providing them with ideal overwintering quarters.

SCARING DEVICES

Some of the larger pests such as cats and birds, can be prevented from damaging the crop by scaring them away. The problem is that the pests gradually get used to some scaring devices and they then begin to lose their effectiveness.

'Humming line' can be bought as a proprietary product, or cassette tape can be used from a worn-out or broken audio or video cassette. It must be stretched taut between canes a short distance above the crop; the wind blows through it to produce a moaning or humming noise. This has been found to be quite effective at scaring birds, but it can also be irritating to neighbours!

Children's windmills, the brighter coloured the better, can be placed strategically in the garden; the rattling and flashing effect can be reasonably effective, but rather larger windmills than the standard size are necessary for the best results. This is one type of scarer that birds quickly learn to ignore. Most birds are upset by hawks, and various devices have been produced which mimic some aspect of predatory birds. A plastic hawk shape on top of a thin whippy pole moves about in the breeze and looks like a hovering bird: its effectiveness seems to vary, but again, birds quickly get used to it and learn that it isn't a threat. Other devices mimic the eyes of a hawk; some glitter to produce a more alarming effect.

ROTATION

Most vegetable crops should be grown in rotation – in other words, moving the crops round from year to year so that the same vegetable (or type of vegetable) does not occupy the same piece of ground in succeeding years. While a crop is growing, pests and diseases particular to that crop will build up in the soil. If the same crop is grown in the same place the following year, these pests and diseases will already be in residence, ready to attack, whereas if a completely new type of vegetable is grown there, the specific pests and diseases should die out.

Table 1. Rotation chart

	Plot A	Plot B	Plot C
Year 1	Add lime	No lime or manure	Add manure
	Brassicas	Potatoes and roots	Peas and beans
Year 2	No lime or manure	Add manure	Add lime
	Potatoes and roots	Peas and beans	Brassicas
Year 3	Add manure	Add lime	No lime or manure
	Peas and beans	Brassicas	Potatoes and roots

Rotation also means that you can take account of specific soil and fertilizer requirements of the different groups of crops. The commonest rotation system splits crops into three groups: brassicas (all members of the cabbage family); root crops (including potatoes); and peas, beans and others.

BRASSICAS

These need lime because it helps prevent the fungus disease clubroot, which likes acid conditions. Adding fresh manure or compost causes the soil to be 'puffy' and light, which results in poor-hearted cabbages and loose, blown sprout buttons; you can add very well-rotted manure and compost the preceding autumn if you like.

Brassicas include Brussels sprouts, cabbages, kohl rabi, cauliflowers, broccoli and calabrese, kale, swede, turnips and radishes.

ROOT CROPS

These must not be grown in freshly manured soil

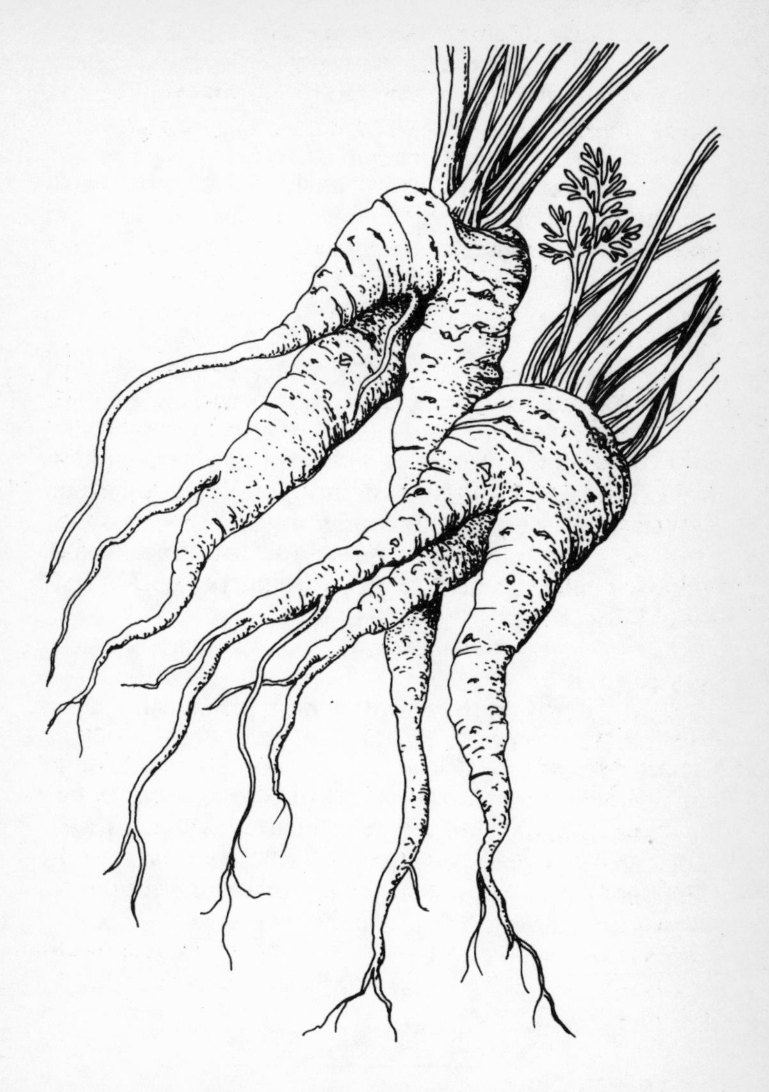

Fig. 5 *Carrots grown in soil containing fresh manure or compost will become badly forked or 'fanged'.*

because they will not grow straight: the fibrous nature of the manure or compost makes the roots fork and fang (Fig. 5). The disease potato scab thrives in alkaline soil, so no lime is added to this plot.

Root crops include beetroot, carrots, parsnips, salsify and scorzonera as well as potatoes. (Root crops which are members of the cabbage family, like swede and turnips, are included in the brassica plot.)

PEAS AND BEANS

These need a moisture-retentive soil, so add plenty of organic matter to this plot. Naturally-occurring nodules on the roots of these plants convert nitrogen in the atmosphere into soil nitrogen which will be available to the following crops of brassicas – greedy plants which will do well in the fertile conditions.

Crops include ordinary peas, snap and sugar peas, runner beans, french beans, broad beans. Onions should also be grown on this plot.

Other crops such as sweet corn, marrows and courgettes, celery, lettuce and spinach can usually be fitted in where required as they do not normally suffer from soil-borne pests and diseases.

Some juggling may be required to fill even-sized plots for these rotations, but it is worth trying to stick to such a scheme where possible. It will cut down the occurrence of pests and diseases, and make the crops easier to look after as similar types are grouped together.

SPACE TO BREATHE

Plants that are grouped closely together produce a humid microclimate where fungus diseases flourish.

Seedlings are particularly prone to damping-off disease when overcrowded, and this can destroy an entire box of seedlings in no time. Sow seeds thinly, and thin them out or prick them out at wider spacing as soon as they can easily be handled. Larger plants, too, should be given enough room to reach their natural proportions, not made weak and badly shaped because they were struggling for light and space.

An important part of fruit pruning is the thinning out of branches so there are none overlapping, rubbing against each other, crowding each other out. Scourges of soft fruit like American gooseberry mildew can often be avoided by keeping an open, airy centre to the bushes.

RAISING YOUR OWN PLANTS

It is often possible to import pests and diseases into your garden on the roots and foliage of plants, whether they are bought from a garden centre or given to you by friends or neighbours. It is a good policy to check plants very carefully before buying them, particularly looking on the undersides of leaves for pests, and choosing the healthiest, strongest-looking specimens. If you are given plants, tactfully ensure these, too, are healthy before bringing them near your garden.

One of the most important diseases which can be imported in this way is clubroot, a fungus disease of brassicas. It lives in the soil and affects the roots of all members of the cabbage family, causing them to become distorted and leading to poor growth. It can exist in the soil for many, many years, lying dormant quite happily if no brassicas are grown. Once it is in your soil you cannot realistically expect to get rid of it, though you can alleviate its effects to some extent.

Obviously, then, you should try to keep clubroot out

of your vegetable garden at all costs, as it can very severely reduce the yields of the many brassica crops commonly grown. As it is impossible to tell whether seedlings and young plants are infected, the safest course is to raise all your own brassica plants from seed. (Although it is possible for seed to carry clubroot, it is not common for it to be brought into gardens in this way.)

Seedlings are particularly vulnerable to all sorts of diseases and soil pests, so when raising your own plants, use a sterile compost rather than garden soil. Once the seedlings are large enough to be planted out they should have sufficient resistance to overcome the harmful organisms that would have finished them off in their infancy. If you don't want to buy proprietary composts you can sterilize your own soil by steaming it, or more conveniently by putting it in a microwave oven for a few minutes. Ensure that pots and seed trays are washed frequently, too, and given a dip in a general purpose disinfectant or bleach before the main seed-sowing season begins.

BARRIERS

Many pest and disease organisms are too small to be stopped by physical barriers, but others are deterred very successfully. There are plenty of methods you can try.

FENCES

Can protect plants from pests as diverse as deer and carrot root fly, though deer proofing a garden will cost a great deal of money. Deer can leap to astonishing heights, and a fence needs to be at least 2.5 m (8 ft) high to keep them out. Visiting dogs can be deterred by much lower fences, though cats will find their way over most barriers if they feel so inclined.

Rabbits can also be kept out of gardens by fencing, but in this case the fence needs to extend below ground as well as above (Fig. 6). Bend up the bottom 30 cm (12 in) of a 1.5 m (5 ft) wire fence and bury this in the ground so that it forms a loop of fencing (facing away

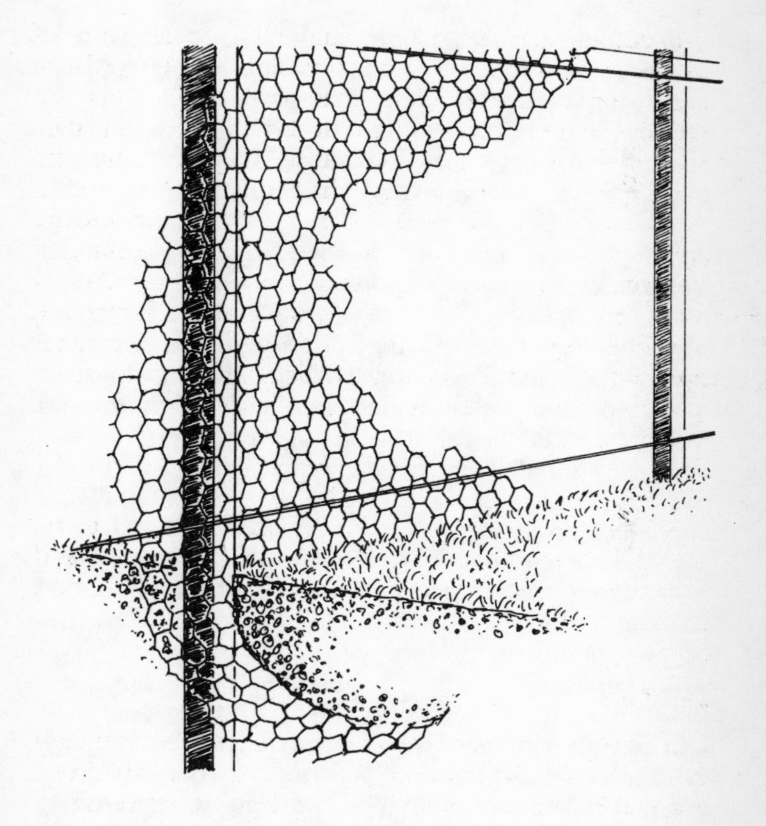

Fig. 6 *Rabbit fencing needs to extend in a loop below ground to prevent rabbits digging their way in.*

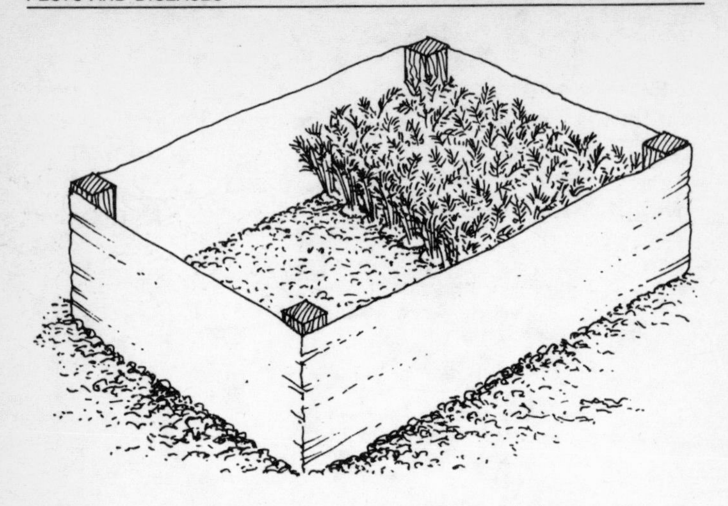

Fig. 7 *Carrot root fly can be deterred simply by erecting a low-level barrier round the plants.*

from the garden) below the soil. When rabbits run up against the wire fence they will start to dig down to work their way below it but your loop below ground will prevent them. Rabbits can also jump well, so you will need some height to the fence; hares jump even better, so make the fence some 20 cm (18 in) taller if they are a problem.

Not many gardeners are troubled by badgers, but if your garden forms part of their regular run you could find them causing damage. They may scrape up patches of lawns and borders looking for food, trample plants and sometimes sample vegetable and fruit crops. They are strong animals and very difficult to deter from their set runs, though heavy duty fencing will sometimes turn them aside. It will be more successful if you provide an alternative route for them to rejoin their established run as soon as possible, though you may not be able to persuade them to use it. You may prefer to relocate plants that are being damaged, provide the badgers with

Fig. 8 *Surround the bases of cabbage plants with discs made from carpet felt or similar material to prevent rootfly attack.*

food at set points and enjoy their company in the evenings: they can become relatively tame.

Carrot root fly sounds an unlikely creature to be deterred by a fence, but because it flies quite close to the ground searching out carrot plants largely by smell, a polythene fence about 75 cm (2 ft) high can keep it away from your crop (Fig. 7). Grow your carrots in blocks rather than rows, as this makes fencing easier. Slugs can also be deterred by this sort of fence; use polythene from fertilizer sacks or similar, cut into strips and nailed to a wooden framework.

NETTING

Can be used instead of more expensive fencing to protect plants from birds, in particular. A walk-in fruit cage is ideal: a wooden or metal framework is covered with netting around the sides and on the top, to keep out birds and other pests. While fruit is usually the crop most at risk, vulnerable vegetables can be grown inside a fruit cage too: cabbage plants often need protection from pigeons in the winter, for example. However, the roof netting should be replaced with a wider mesh netting for the winter in case there is a fall of snow. Small mesh netting will hold the snow, and the weight will damage the supports and tear the netting. In late winter many gardeners like to remove all the roof netting to allow birds free access to the fruit bushes – they will eat a wide range of pests that have overwintered on the plants.

INDIVIDUAL PROTECTION

Can be given to some plants. Brassica plants, for example, can be attacked by cabbage root fly, which lays its eggs in the soil near the plant stems. Surround each plant with a collar – purpose made ones can be bought, or you can cut your own from carpet underlay or heavy card – and no eggs will be laid (Fig. 8). Young trees can be killed by rabbits, hares and squirrels stripping the bark from the trunk: spiral plastic tree guards should be fitted round all newly planted trees. An even better start can be given by providing the tree with a corrugated plastic shelter which will protect it from adverse weather as well as a range of pests.

CLOCHES

These come in a range of shapes and sizes. Originally they were intended to give plants protection from the weather, but they protect them from pests, too. Glass cloches are awkward to put together, heavy to move about and can be dangerous, especially where there are children; the glass also soon becomes covered with algae which is difficult to remove. Plastic cloches are cheaper, but they are blown about by high winds and are not so effective at trapping the sun's heat. Both types are good at protecting plants from birds and other pest mammals such as rabbits. Because they are like a mini greenhouse, the plants grow quite lush and 'soft' under

Fig. 9 *Empty lemonade bottles make good individual cloches if the bases are cut off.*

a cloche, which can make them more prone to disease attack.

A very simple cloche can be made from a plastic lemonade bottle (Fig. 9): simply cut the bottle in half and use the halves to place over young plants to protect them in their early stages. Leave the lid off the top half to provide some ventilation, and make some holes in the bottom half for the same purpose. These bottles can also be cut into wide rings to put round plants that are particularly vulnerable to slugs, and will cut down the amount of damage they suffer.

A relatively new type of cloche is the floating cloche or horticultural fleece. These are made from perforated plastic or a fibrous material called spun polypropylene: they are laid over the young crop (usually vegetables) and can be left in place until harvest time. They protect plants from frost and allow a favourable microclimate to develop so that plants grow rapidly; water and air can pass through the material without any problem. They also protect plants from a wide range of pests and, to a lesser extent, diseases. If treated carefully, the cloches can be used many times.

One of the drawbacks to these types of cloches is that weeds tend to develop unchecked beneath their cover: also it is very difficult to see what is happening to the crop. Although the material will stretch and expand, it can sometimes damage plants and cause distorted growth.

PUTTING THEM OFF THE SCENT

Because some pests seek out their target plants by scent, it is possible to put them off by masking the scent in some way. This may be by simply making sure you avoid the plants giving off a strong scent. Bruising the foliage of carrots, for example, will attract carrot root fly: sow carrot seed very thinly to avoid the need for thinning the seedlings, and weed or thin rows of carrots at dusk to give the fly minimum time to find them. You can also grow strong-smelling crops near one another – carrots and onions, for example, will tend to cancel out each other's scents. Some gardeners grow strong-smelling herbs like tansy round the edges of the vegetable garden to put off scent-programmed pests.

This leads us on to 'companion planting', which has quite a following among some gardeners. The theory is that certain plants grow particularly well together, and will protect each other from disease. Garlic is said to prevent rose blackspot if grown beneath rose bushes, or peach leaf curl when planted under peach trees; while certain flowers and vegetables are thought to have a natural antipathy to each other and should never be grown near each other or they will not thrive. However, there does not seem to be any scientific support for any of these theories except that pests who rely on scent to find their targets may be put off by more strongly scented plants near by.

Sowing times can be altered to avoid the worst times for pest and disease attack on certain crops. Broad beans are very often attacked by black aphids, which home in on the growing tips. If seed is sown in the autumn, by the time aphids are ready to attack, the broad bean plants are well developed; their growing tips are not so tender as spring-sown plants so are less attractive to aphids, and plants are large enough to have the growing tips pinched out if they are affected. Peas can be ruined by pea moth, but early or late-sown peas are much less likely to be affected than main-season sowings.

If you have a greenhouse, it is often better to start off plants in there, be they vegetables or flowers, and plant them outside when they are better developed and more able to withstand pests or diseases than newly germinated seedlings would be. If you have clubroot in your soil, you can still achieve reasonable crops of brassicas by sowing seed in sterile compost in the greenhouse and pricking out seedlings into individual pots of compost before setting them out in the garden.

TRAPS

If you can't prevent the invaders coming into your garden, you can at least try to waylay them before they reach your plants. The types of pest traps available to gardeners are legion.

Probably everyone has tried hanging the sticky jam jar half full of water in a plum tree, in an attempt to entice wasps away from the plums, and to their death by drowning. It is reasonably successful in that plenty of wasps seem to drown, though whether the plums suffer any less damage is debatable.

SLUG TRAPS

These operate on a similar principle as sticky jam jars. Shallow dishes half filled with beer (for some reason slugs and snails appear to like beer) are set in the soil round plants that are particularly vulnerable to slug damage, with the lip of the dish level with the soil. Slugs are attracted to the beer and in an intoxicated state are unable to crawl out again, so they drown – at least, that's the theory. Once the beer has done its work on a number of slugs, however, future visitors are able to crawl out over the bodies of their dead friends, so dishes must be emptied and refilled regularly if they are to keep working. Another drawback is that they can also attract and destroy helpful creatures such as ground beetles.

Other types of slug traps rely on chemical slug baits to kill the slugs, but are designed to make the slug bait less likely to be accidentally eaten by birds, pets or hedgehogs. Half grapefruit skins are also used: the slugs are attracted to the moist interiors as a nice cool place to spend the day, and the gardener can make regular rounds to empty the skins and dispose of the contents.

FLY PAPERS

These are another old-fashioned remedy. Strips of paper coated with an intensely sticky substance are hung up in kitchens or other places where flies are a nuisance; as soon as a fly alights on the paper or brushes past it it is stuck fast and dies a rather slow death by starvation and exhaustion. Only slightly more sophisticated are whitefly traps available for greenhouses; they are coloured yellow in an attempt to attract whitefly to them and may come with a protective grate which prevents plant foliage (or the gardener's hair or clothes) coming into contact with them. Giving infested plants a light shake

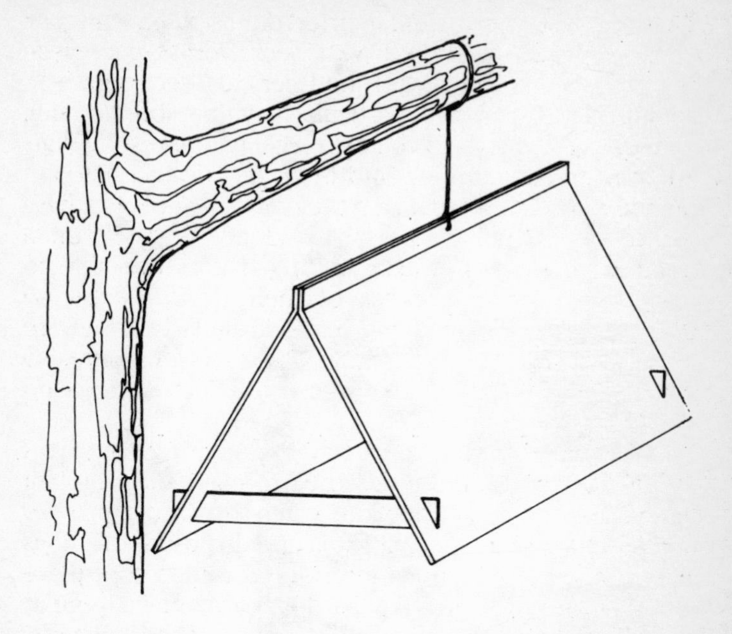

Fig. 10 *Male codling moths are attracted to sticky traps by the use of pheromones.*

will send clouds of whitefly into the air and increase the chance of some meeting their end on the sticky strips.

CODLING MOTH TRAPS

These are more cunning and use pheromones – chemical sex attractants given off by the female moths to lure the males. Traps shaped like a house roof are smeared on the inside with the pheromone and glue and hung in the branches of apple trees; male moths are

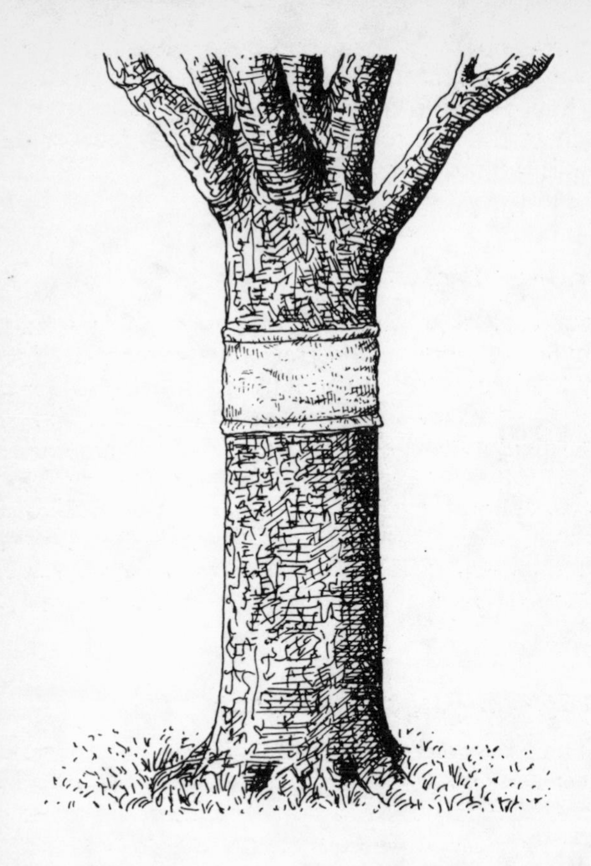

Fig. 11 *Grease banding fruit trees helps prevent the wingless female winter moths climbing up the trunks.*

attracted by the pheromones and stick to the glue (Fig. 10). These were originally used in commercial orchards not as a control, but to judge when there were sufficient moths flying to make using a pesticide worthwhile. In a garden situation they may be sufficient to cut down damage on their own, but once the sticky surface is covered with moths it becomes ineffective. Another style of trap available also uses pheromones to attract the male moths, but relies on water in the base of the trap to kill the moths by drowning.

GREASE BANDING

Another serious pest of apples is the winter moth; the female is flightless, and crawls up the trunks of trees to lay her eggs. Grease banding the tree will catch her on the way up. Strips of greaseproof paper are coated with the same type of sticky glue found in the fly and codling moth traps: they are tied tightly round the trunks of trees, about 1 m (3 ft) from the ground (Fig. 11). The glue itself can instead be spread directly on the tree trunk, which is more effective on rough-barked trees where moths may be able to crawl beneath the paper.

ELECTRIC INSECT KILLERS

Butchers' shops these days all seem to display large electric insect killers with a powerful ultra-violet light to attract the insects to a sudden death. These are quite expensive pieces of equipment, but smaller models are available for home use. They work effectively as long as the insects can be persuaded near enough to the trap to be electrocuted: different insects have different responses to ultra-violet, but flies do not seem to be particularly attracted by it. Wasps and various midges, on the other hand, are killed by the trap, but as far as I know, not much work has been done on assessing its suitability as a plant pest killer. In theory, a pheromone lure could be used to make this trap suitable for an enclosed space such as a greenhouse.

WEEDS

A weed, as the well-known saying goes, is simply a plant in the wrong place. Generally weeds are wild plants, germinating from seeds already present in the soil, or ones that have been brought in by birds or in the soil of new plants. Weeds are not always wild plants – cultivated plants can easily become weeds when they grow too well or turn up in the wrong part of the garden. Potatoes may be a flourishing and valuable crop one year: next year, the 'volunteer' plants arising from overlooked tubers will not be at all welcome amongst the peas.

WHY WEEDS MATTER

Weeds compete with crop plants for nutrients, water, light and space, and they often win. Crops can be smothered by weed growth, and even if they appear to be holding their own, their vigour and subsequent yields are much reduced. In the ornamental garden, weeds detract from the appearance of the garden plants. If allowed to set seed, weeds continue to be a problem for years to come.

On the other hand, some people believe that weed growth gives shelter to crop plants and they will do better if left unweeded. Weeds also give shelter to several garden pests – and to useful predators, too. The case against weeds is far from clear cut, but most gardeners find a weedy garden irritating and unattractive. Some areas of 'wild flowers' can be left to provide refuge for predators in an appropriate area of the garden.

TYPES OF WEEDS

Annual weeds. As their name suggests, these die

after flowering, usually completing their life cycle in one season. Their growth can be very rapid, but they are not generally as big a problem as perennial weeds, and most are fairly easy to control. They reproduce themselves by scattering seed prolifically (hence the saying 'one year's seeding, seven years' weeding') and should always be removed before they have time to flower and seed if at all possible.

Perennial weeds. These continue to grow from year to year, and are not killed by the winter (though many of them disappear from sight during the winter months). They may also seed themselves, though not usually as freely as annual weeds. They are more of a problem to control than annual weeds and require persistence, particularly when using non-chemical methods.

CONTROL

Chemical weedkillers work in a variety of complex ways. Non-chemical means of weed control can be just as effective, as easy to carry out and a lot cheaper. As with most organic pest control methods, prevention is usually easier and more successful than cure.

Mulching. Weeds need light in order to grow, just as garden plants do: mulching helps to deprive them of light and so prevent their growth. Before applying a mulching material, the bed should be cleared of all visible weeds and the soil should be thoroughly moist.

Black polythene is one of the most effective weed-killing mulches. It can be spread over a piece of ground which is to lie fallow for a year and should kill most weeds that attempt to grow under it. Anchor it firmly by burying the edges in the soil, and scatter some soil or similar material over the surface both to give it a more pleasing appearance and help it stay in place. If you wish to grow plants in the ground being treated, plant through slits cut in the polythene.

Other mulching materials include old newspapers – one good way to recycle them – grass clippings, garden compost, straw, forest bark chippings, and old carpets: virtually anything that will cover the soil to keep moisture in and light out. Bark chippings are quite expensive but look very attractive in an ornamental garden. Unrotted organic material like straw and grass clippings will eventually add nutrients to the soil, but in their unrotted state will first take nitrogen *out* of the soil while decomposing, so don't overdo their use on hungry crops. Sheets of newspaper need soaking and then burying lightly or they will soon be distributed over the entire garden.

Ground cover. Like mulching, ground cover relies on covering bare soil to prevent weeds colonizing it – only in this case you cover the soil with plants. This is only useful in the ornamental garden. There are several plants that form good ground cover, though some of them are invasive and can end up becoming weeds themselves. Ground cover plants are usually grown beneath shrubs or trees which can withstand the extra competition. Obviously the plants will compete for food and water just as weeds would, but the attractive appearance of a border is maintained without the work of weeding.

Table 2 Good ground cover plants

Ajuga reptans	Carpet forming, multi-coloured leaves. Low.
Alchemilla mollis	Rounded, crimped leaves, yellow feathery flower heads. Medium.
***Bergenia**	Large, round, glossy leaves, pinkish flowers. Medium.

***Calluna (ling)**	All-year interest from a range of varieties, both flowers and foliage colour. Acid soil required. Low–medium.
Cornus canadensis	Rounded leaves in groups of four, turning red in autumn. White flowers and red berries. Medium.
***Cotoneaster**	Several suitable varieties. Small leaves, red berries. Low–medium.
***Erica (heather)**	As calluna. Mainly like acid soil but some lime-tolerant types available. Low–medium.
***Euonymus fortunei**	Several suitable varieties. Small leaves, some variegated. Good year-round interest.
***Gaultheria**	Small, dark green, glossy leaves. White flowers followed by red berries. Vigorous. Medium.
Geranium	Soft, mid-green leaves and blue or pink flowers. Several suitable varieties. Medium.
***Hebe**	Several suitable varieties. Dark, shiny elliptical leaves and bottle-brush type flowers in red, pink, blue or white. Medium.
Hedera	Vigorous, scrambling plants; many good varieties with wide range of leaf shape, size and colour. Low.
Hosta	Striking, often 'sculptured' leaves of green, blue or variegated. Bell-like white or pale lilac flowers. (Prone to slug damage.) Medium.
Hypericum	Light green, lanceolate leaves and large yellow saucer-shaped flowers. Very vigorous. Medium.
***Juniperus**	Spreading conifer, blue green or various shades according to variety. Low.
Lamium	Silvery variegated nettle-like leaves and white or pink flowers. Quick growing. Low–medium.
Polygonum	Several suitable varieties. Quick to spread, with pink 'pokers' of flowers. Low–medium.
Roses	Several prostrate varieties available, with flowers in a range of colours. Low.
***Sarcococca**	Dark green, narrow, glossy leaves and white winter flowers. Vigorous. Medium.
***Saxifraga umbrosa**	Rosettes of mid-green leaves and clouds of fluffy, light pink flowers in early summer. Medium.

Thymus	Many varieties with small, aromatic leaves and pink or white flowers. Low.
***Vinca**	Dark green, glossy, lanceolate leaves on creeping stems; blue, purple or white flowers.

* = evergreen

Hoeing. One of the best ways of dealing with annual weeds. A hoe should have a very sharp blade and be pushed along the soil surface to sever weeds from their roots; it is not meant to be used to dig weeds up, as many gardeners believe. Although the roots are left in the soil they seldom regrow as their food factory – the green part of the plant – has been removed. If they do regrow, repeated hoeing will weaken and eventually kill them.

Hoeing should be done while weeds are very small to be most effective. Choose a dry, bright or windy day for preference, then any weeds that come up with their roots will soon wilt and die whereas in cool, moist weather they could re-establish themselves. Hoeing can also be used to weaken and possibly kill perennial weeds, though more persistence is usually needed here.

Flame gun. A dramatic way of dealing with weeds! A paraffin-powered flame gun will literally scorch off their topgrowth and can also destroy the roots, depending how long it is allowed to play on the soil. The drawback of this method is that it also kills useful soil-living organisms so should be used with caution. It is particularly useful for hard surfaces such as gravel paths. The heat generated is intense, so be careful when using it on paving – it can crack the slabs. You should also be aware of the risk of fire in dry weather.

RECOGNIZING AND TREATING SOME COMMON WEEDS

Bindweed. Rapid-growing, twining plants with arrow-shaped leaves and trumpet-shaped pink and white flowers. The fleshy underground stems snap easily and each piece left in the soil sends up new shoots. Plants can be forked out carefully, but this is a very difficult weed to eradicate. Badly infested land can be covered with black polythene for a season.

Buttercup. The three-lobed, toothed leaves are lightly hairy; the well-known cup-shaped flowers bright yellow. Prise out plants with a hand fork. Creeping buttercup, which forms long chains of plantlets on runners, is particularly difficult to control and you will need persistence.

Lesser celandine. This plant forms rosettes of heart-shaped, shiny green leaves on long stems. Glossy yellow flowers are similar to buttercups but with narrower petals; they appear very early in the year. Celandine spreads very quickly on creeping stems. Bulbils are sometimes found in the leaf axils, and these fall to the ground to produce new plants when the weed is pulled up or hoed off; persistence is necessary to achieve control.

Common chickweed. Sprawling stems bear soft, pale green leaves and small, white flowers which appear all through the year. Chickweed spreads rapidly, soon forming a bright green mat if allowed to. It is easy to hoe off, however, or it can be grubbed up by hand between plants. Try to remove it before it seeds.

Couch grass. Couch looks like a slightly hairy, rough, clump-forming grass. The tough, wiry roots of this plant soon form a mat below the soil surface. Pieces of roots which are broken off produce new plants, making this a very difficult weed to control, particularly where it occurs amongst cultivated plants. On vacant ground dig out as much root as you can and cover the soil with black polythene for a season to try to starve the couch grass out; among plants keep pulling out as much as possible.

Daisy. Low rosettes of toothed leaves and many-petalled, white, pink-tipped flowers with a yellow centre – daisies are very well known, particularly on lawns. Dig up plants with a trowel or hand fork. Many people like the appearance of a daisy-studded lawn and are content to let them stay.

Dandelion. A carrot-like root holds this common weed firmly in the soil. Jagged-edged leaves are topped by the familiar yellow, many petalled flowers which turn to fluffy white 'clocks' of seeds by which the plant spreads widely. Because the rosette of leaves is untouched by the lawn mower, dandelions do well on lawns. Like daisies, dandelions can be dug out with a fork or trowel but take care not to break the brittle root.

Dock. These tall plants have large, coarse-looking oval leaves and spikes of brownish flowers. Docks are common weeds of poor soil. They have a deep taproot which must be dug out carefully with a fork or spade. Improve the soil and check drainage to prevent a recurrence.

Fat hen. This weed grows on fertile, humus-rich soil. The toothed, lanceolate leaves have a distinctive mealy appearance, especially on the undersides. Plants are not difficult to control – pull them up or hoe them off when young. They tend to appear mainly in vegetable plots.

Ground elder. A low-growing, rapidly spreading weed with bright green, toothed leaves and insignificant white flowers in early summer. The underground stems spread very rapidly and form a dense mat which is very difficult to eradicate. Dig up as much of the plant as you can. Repeated hoeing of the topgrowth will weaken and eventually kill it; The leaves are edible so it has some uses!

Hairy bitter cress. Small, round, dark green leaves, are carried in opposite pairs up the stems, which form a tight rosette. Flower stems are tall and slender and carry insignificant white flowers which turn to long green seed pods. The seed pods explode when touched, spreading the seed widely. This weed is particularly common in garden centres, growing in the compost of container-grown plants. Hoe it off when young, choosing a dry day, and keep hoeing the new generations of seedlings as they germinate.

Horsetail. This weed, which looks like a miniature Christmas tree, grows on poor soils and is extremely difficult to eradicate due to its wiry underground runners. Cut off or pull up the topgrowth frequently in order to weaken the plant, but covering infested ground with black polythene is probably the best way to deal with it. Improve the soil to prevent a recurrence.

Japanese knotweed. Not very common, but a spectacular problem where it does occur. The bamboo-like stems grow rapidly up to 3 m (10 ft) tall; large, heart-shaped leaves are carried alternately up the stems. Panicles of small white flowers occur in the leaf axils. Underground stems mean the plant spreads quickly and can form a dense thicket if not dealt with promptly. Use a mattock to remove and weaken as much of the rootstock as possible, but covering an infested area with black polythene is probably the best treatment. There is some hope of a biological control for this weed; in Japan it is not so much of a problem as natural predators keep it under control.

Oxalis. These plants reproduce very successfully from tiny bulbils round the rootstock; cultivated varieties often get out of hand and become pernicious weeds. The leaves are clover-like and the flowers pinkish trumpets on a long stem. Dig up plants carefully in very early spring (before the bulbils form to avoid spreading them) or cover areas with black polythene to starve the plants out. Another difficult weed to control.

Shepherd's purse. A very common weed with rosettes of ragged, toothed leaves. Small white flowers are followed by heart-shaped seed pods which give the plant its name. Several generations are produced each season. Shepherd's purse is quite easy to control by hand pulling and hoeing, but you will need to persist with the new generations of seedlings.

Stinging nettle. A very well-known plant, with toothed, deep green, lance-shaped leaves with stinging hairs. Nettles are a major food plant for butterfly larvae, so if you can, leave them to grow. If they have to be removed, pull them up (wearing gloves to avoid painful stings) or use a fork on larger plants.

CHAPTER 4

CONTROLLING PESTS AND DISEASES

Despite the preventive measures you may take, there will be times when pests and diseases strike your garden. This need not be a disaster as by taking early action, you may be able to prevent much damage being done. Always be observant, looking out for the first signs of attack so that you can deal with it straight away.

If you have been used to using chemical controls, expect damage to be worse than normal for the first season after you stop spraying; it will take some time for the population of natural predators to increase sufficiently to deal with the pests. A balance will eventually be achieved, however, so don't despair.

When you do spot a pest or disease, the measures you can take against it are usually fairly straightforward. Sometimes there is not much that can be done, short of removing affected plants promptly so that the problem doesn't spread. In other cases, there may be a very simple and effective remedy. Occasionally, the problem has little effect on the plant and there is actually nothing that needs to be done about it.

The following charts list some of the reasonably common troubles that you may come across in your garden. The numbers at the end of each entry refer to the table of control measures on pages 77–79, which tells you what action you can take.

Although the number of potential problems may look alarming, they are not all serious – and you are unlikely to come across more than a small percentage of them.

RECOGNIZING AND TREATING COMMON PESTS AND DISEASES

FRUIT

Apple and pear

Aphids: leaves and tips of shoots sticky and covered with small insects. 1,2,3.
Apple sawfly: fruits misshapen with scars on surface. 2,4.
Bitter pit: fruits with small black bitter spots under skin. 5.
Capsids: small holes in leaves. 2.
Caterpillars: larger holes in leaves. 2,6.
Codling moth: small caterpillars tunnelling in fruit with holes from core to outside. 2,16,15.
Fireblight: shoots die back from tips with scorched leaves hanging on. 7.
Fruit tree red spider mite: leaves flecked and bronzy. 2,8.
Honey fungus: tree dies back gradually from branch ends, orange toadstools present at base of trunk. 9.
Mildew: leaves and shoot tips covered with white powder. 10,11.
Pear and cherry slugworm: leaf surfaces eaten; small, black slug-like creatures present. 12.
Scab: rough, raised, corky patches on surface of fruit. 13.
Tortrix moth: small caterpillars tunnelling in fruit near surface. 6.
Winter moths: small, 'looping' caterpillars; flowers and buds with small holes. 14.
Woolly aphids: cluster of white, cotton-wool-type substance on branches and trunk. 1,2,3.

Cherry
Birds: fruits eaten. 16.
Cherry blackfly: shoot tips and foliage heavily infested by small black insects. 1,2,3.
Honey fungus: (see under apple).
Silver leaf: leaves appear silvery, branches die back. 10,13.

Currants and gooseberries
American gooseberry mildew: leaves and fruits covered with white mould turning to brown 'felt'. 10,11,13.
Aphids: leaves blistered and puckered; small insects present. 1,2,3.
Big bud mite: swollen buds in spring (Fig. 12); poor crops. 10.
Gooseberry sawfly: leaves eaten, often only a network of veins left. 2,17.

Peach
Peach leaf curl: leaves puckered and red, falling early 13.
Shot hole: leaves with small holes. 18.
Silver leaf: (see under cherry).
Virus: leaves streaked, poor growth. 10.

Plums
Aphids: leaves twisted and distorted, especially at shoot tips; small insects present. 1,2,3.
Brown rot: fruits rotted, sometimes with white pustules. 10,13.
Sawfly: fruits with tunnels made by small caterpillars. 2.
Silver leaf: leaves with a silvery sheen; branches dying back. 7.
Wasps: cavities eaten into fruits when ripe. 19,20.

Raspberries and other cane fruits
Botrytis: fruits rotted, with powdery grey mould. 19.

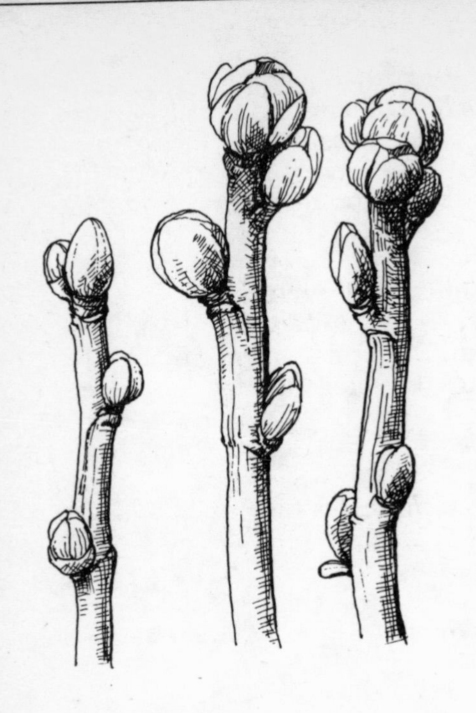

Fig. 12 *Big bud mite on blackcurrants can be identified by the presence of swollen buds in winter.*

Cane and leaf spot: purplish blotches on canes and leaves; poor performance. 10.
Raspberry beetle: maggots in fruit. 2.
Virus: leaves streaked and mottled; performance poor. 10,13.

Strawberries
Aphids: leaves puckered and malformed; clusters of small insects. 1,2,3.
Birds: fruits pecked or disappearing completely. 16.
Botrytis: fruits rotted with grey mouldy growth. 10.

Fig. 13 *Broad beans are frequently attacked by aphids (blackfly) which cluster near the tips.*

Frost damage: flowers appear black in centres and do not set fruit. 18.
Mildew: leaves with powdery white coating. 13.
Slugs and snails: fruits partially eaten. 21.
Viruses: leaves streaked and mottled; performance poor. 10.

VEGETABLES

Beans

Black bean aphid: (Fig. 13) leaves and shoots thickly infested by small black insects. 1,2,3.

Chocolate spot: dark brown spots on all parts of broad bean plants. 22.
Halo blight: round, brown spots on leaves, surrounded by a yellow 'halo': bacterial. 22.

Beet
Downy mildew: pale spots on leaves with mould growth beneath. 22.
Leaf miner: tunnels in foliage. 23.
Scab: corky, raised patches on roots. 24.

Brassicas
Cabbage root fly: plants wilt rapidly in warm weather; roots infested by white maggots. 25.
Cabbage caterpillars: foliage extensively eaten; caterpillars present. 6,17.
Clubroot: plants fail to thrive; roots swollen and distorted. 26.
Downy mildew: foliage with brown patches and grey mould on undersides. 10,22.
Mealy cabbage aphids: undersides of leaves infested by small, mealy, grey insects. 1,2,3.
Powdery mildew: pale patches on leaves with mould growth below. 10,22.
Whitefly: small, white, moth-like insects sometimes accompanied by black sooty mould. 18.

Carrots
Carrot-willow aphid: small green insects clustered on foliage. 1,2,3.
Carrot fly: foliage has reddish tinge; roots tunnelled by larvae. 27.

Lettuce
Aphids: foliage infested by small green insects. 1,2,3.
Botrytis: fluffy grey mould present on foliage and/or stems. 28.

Caterpillars: holes eaten in foliage. 6,17.
Downy mildew: foliage with yellow patches on surface and fluffy white mould beneath. 22,28.
Root aphids: plants fail to thrive; roots infested with small grey insects. 10,30.
Slugs and snails: holes eaten in foliage, particularly on young plants; slime trails evident. 17.

Marrows and courgettes
Botrytis: fruits rotting with grey fluffy mould starting at flower end. 21.
Mosaic virus: foliage mottled and plants unthrifty; fruits misshapen. 10.
Powdery mildew: leaves with white powdery mould growth. 28.

Onions (including leeks, shallots and garlic)
Neck rot: bulbs (usually in store) rot from neck downwards.
Onion fly: leaves yellow; maggots present in bulbs and stems. 10.
White rot: bulbs and stem bases covered in white fluffy mould with small black spots. 10.

Parsnip
Canker: brown rot around shoulders of root. 30.
Carrot fly: plants unthrifty; white maggots present in roots. 27.
Celery fly: mines in foliage. 10.

Peas
Birds: peas taken from pods. 16.
Downy mildew: pale patches on leaf surface with mould growth beneath. 22.
Pea moth: small caterpillars inside pods, eating peas. 31.
Pean and bean weevil: notches eaten from the edges of leaves. 13.

Powdery mildew: leaves, stems and pods with powdery white patches. 13,22.

Potato

Aphids: leaves and shoot tips infested by small green or pink insects. 1,2,3.
Blight: leaves with black spots, foliage quickly becoming yellow and falling. 32.
Slugs: larger holes eaten in tubers, often with secondary rotting. 33.
Virus: leaves distorted and/or streaked and mottled. 10,34.
Wireworm: small round tunnels made in tubers. 23.
Scab: raised, corky patches on tubers. 24,30.

Spinach

Beet leaf miner: foliage with brown mines. 10.
Downy mildew foliage with yellow patches and mould growth beneath. 28.

Tomato

Blight: fruit and foliage with dark, blotchy rot (outdoor tomatoes). 10.
Blossom end rot: hard, shrunken brown area at blossom end of fruit. 5,35.
Bortytis: fluffy grey mould on damaged parts of plants spreading to healthy areas. 28.
Leaf mould: foliage with yellow patches above and grey mould growth beneath. 10,13.
Viruses: foliage streaked and/or mottled and distorted. 10,34.

ORNAMENTAL PLANTS

Trees and shrubs

Aphids: leaves and shoot tips infested with small, black or green insects. 1,2,3.

Caterpillars: ragged holes eaten in foliage. 6,17.
Clematis wilt: plants collapse and die suddenly. 38.
Coral spot: small, orange spots on dead and (less usually) living wood. 12.
Honey fungus: plants die, producing characteristic orange toadstools. Disease spreads rapidly. 9.
Leaf spots: various blotches and spots on foliage of a wide range of plants and trees. 12.
Root rot: plants fail to thrive; areas of roots rotted. 10,33.

Flowers
Aphids: leaves and shoot tips infested with small, black or green insects. 1,2,3.
Black spot: fringed black spots on foliage of roses; early leaf fall. 10,11,12.
Botrytis: fluffy grey mould, esp. on damaged parts of plants. 10,37.
Caterpillars: ragged holes eaten in foliage. 6,17.
Downy mildew: yellowish patches on leaf surface with fluffy white mould beneath. 10.
Earwigs: holes eaten in leaves; distorted, damaged flowers. 17.
Leaf spots: various blotches and spots on foliage of a wide range of plants and trees. 12.
Leaf miner: tunnels in foliage. 10.
Lily beetle: small, scarlet beetles eating lily foliage. 17.
Narcissus fly: bulbs fail to grow or produce leaves and no flowers; bulbs infested with grubs. 10,41.
Powdery mildew: powdery white mould on leaf surface. 10.
Root rot: plants fail to thrive; areas of roots rotted. 10,33.
Rust: small yellow spots on leaf surface; powdery orange pustules beneath, esp. roses, hollyhocks, antirrhinums, pelargoniums. 10.
Scale insects: small, usually brown, immobile scales on stems and foliage, often accompanied by sooty mould. 42.

Slugs and snails: foliage and stems eaten, esp. seedlings; slime trails evident. 17.

Greenhouse plants

Aphids: leaves and shoot tips infested with small, black or green insects. 1,2,3.
Black leg: black rot at base of cuttings, esp. pelargoniums. 10,36.
Botrytis: fluffy grey mould, esp. on damaged parts of plants. 10,37.
Caterpillars: ragged holes eaten in foliage. 6,17.
Damping off: seedlings collapse at soil level. 39.
Downy mildew: yellowish patches on leaf surface with fluffy white mould beneath. 10.
Earwigs: holes eaten in leaves, distorted, damaged flowers. 17.
Leaf miner: tunnels in foliage. 10.
Mealy bug: small, grey, woolly creatures clustered on stems and foliage of a wide range of plants. 40.
Powdery mildew: powdery white mould on leaf surface. 10.
Red spider mite: speckled foliage with webbing at tips; plants fail to thrive. 6.
Rust: small yellow spots on leaf surface; powdery orange pustules beneath, esp. roses, hollyhocks, antirrhinums, pelargoniums. 10.
Scale insects: small, usually brown, immobile scales on stems and foliage, often accompanied by sooty mould. 42.
Slugs and snails: foliage and stems eaten, esp. seedlings; slime trails evident. 17.
Whiteflies: small, white moth-like creatures, often accompanied by sooty mould. 6.

CONTROL METHODS FOR PESTS AND DISEASES

1. Wash off with a strong jet of water, being careful not to damage foliage.
2. Spray with organically approved insecticide (see pages 84–87).
3. Spray with pirimicarb if you do not mind using a safe but non-organic control.
4. Pick up and destroy fallen fruitlets in early to mid summer as they may harbour the pest.
5. Ensure soil is not too acid (pH 6.5 or more) and/or spray with calcium nitrate or calcium chloride solution.
6. Use a biological control.
7. Cut out affected branches to a point where the wood appears healthy and not discoloured.
8. Natural controls should take over if no treatment is applied.
9. Remove affected plants and as much surrounding soil as possible promptly.
10. Remove and destroy affected plants or parts of plants.
11. Prune to prevent overcrowded branches.
12. Pick off when seen, but no treatment usually necessary.
13. Remove and burn fallen leaves and other debris which may carry the infection over winter.
14. Use grease bands.
15. Use a pheromone trap.
16. Use bird scarers.
17. Hand pick and destroy pest.
18. No treatment.

19. Remove overripe or damaged fruit.
20. Hang a jam jar half filled with sugared water in the tree branches to trap wasps.
21. Keep area around plants open and free from debris: use straw underneath to keep fruits off soil.
22. Choose a more open site for the next crop and do not space plants too closely. Do not save seed of beans affected with halo blight.
23. Cultivate soil to expose overwintering pests to predators.
24. Worst on light, sandy soil: improve water-holding capacity by the addition of organic matter. Do not lime soil.
25. Protect plants with felt (or similar) collars around stems at planting time.
26. Do not buy in plants, which may be infected, but raise your own from seed. Once land is contaminated, raise plants in individual 9 cm (3½ in) pots of compost before planting out.
27. As carrot fly is attracted by the scent of bruised foliage, do not thin or weed unless necessary, then at dusk. Use a polythene barrier around beds.
28. Remove affected bulbs; do not store damaged bulbs.
30. Grow resistant varieties.
31. Avoid sowing in early to mid spring – sow earlier or later than this.
32. Cut down and destroy foliage immediately infection is spotted, to prevent rain spreading the disease to tubers. Remove all tubers from the soil to prevent infection being carried over to next year.
33. Improve soil drainage.
34. Control aphids, which spread virus diseases.

35. Ensure plants do not go short of water while flowering.
36. Use sterilized compost and do¨ot overwater plants.
37. Remove all plant debris and damaged parts of plants promptly; keep greenhouse tidy.
38. Cut off top growth and earth up base of plant to encourage new shoots. Do not plant new clematis in same area.
39. Sow thinly, using sterilized compost.
40. Pick off or treat colonies with small brush dipped in methylated spirits.
41. Pull soil around plants to fill in holes left as leaves die back.
42. Scrape off pests with fingernail where possible.

CHAPTER 5

BIOLOGICAL CONTROL

'Set a thief to catch a thief' the saying goes, and this is what biological control is all about. Research has shown that the overall levels of pest damage in gardens where chemical controls are used and in 'no chemical' gardens is surprisingly similar. This is because pests have their own, natural controlling agents which are always present in the garden and will carry out much of the work for you.

You can take this a stage further by actually introducing some natural controls to pest-infested plants, and this is what is known as biological control.

Some biological control agents are insects, many of which are not native to this country but are especially bred in laboratories for the purposes of pest control. It is difficult to use them on outdoor plants because the controls would either fly away, or would find conditions too cold and exposed to work properly, so biological control achieves most success in greenhouses.

If the controls are very efficient, they will destroy all their prey and die out; you often need to have a low level of pest infestation present all the time to keep the biological controls alive. If the pest species multiplies rapidly, the biological control will do so as well (after a short delay) so it should always keep on top of the problem. It is difficult to keep the populations nicely balanced, so you may need to buy in further batches of controls throughout the growing season.

The control species are often very similar to the pest

species, and will be harmed by most of the pesticides that could be used on the pest. It is possible to find some chemicals that harm the pest and not the control, and these can be used occasionally if the control species seems to be getting swamped by the pest. This is known as integrated control. However, it's best to avoid all pesticides to be on the safe side.

The following is a list of common greenhouse pests and biological control methods, given in order of their popularity with, and availability to, amateur gardners.

RED SPIDER MITE

This common greenhouse pest, together with whitefly, was one of the first to be the subject of widespread biological control methods. The mites are tiny and only just visible to the naked eye. The signs of infestation are a characteristic silvery mottling of the foliage, caused by the sap-sucking feeding habits of the pest, and webbing, like spiders' webs, at the tips of shoots and leaves. If you examine this webbing with a hand lens, you can make out the tiny mites running along it and over the undersides of the leaf. Cucumbers, tomatoes and a variety of house plants are commonly affected, and can be killed by a severe infestation.

The control is a predatory mite called *Phytoseiulus persimilis*. This is similar in appearance to its prey, but is a little larger and more of an orange colour. It is fast moving, feeding on both the adult mites and their eggs.

Plants should be examined regularly and very carefully for the first signs of attack by red spider mite. As soon as it is spotted, the control should be ordered according to greenhouse size. The predators are supplied on pieces of bean leaf, usually with some red spider mites present as food for the journey and while they establish themselves. Open the packages in the greenhouse, and distribute pieces of bean leaf evenly among the infested plants. The predator likes warm conditions so keep the greenhouse at about 25°C (77°F) for the best results, going down to a minimum of 13°C (55°F).

Within three or four days, new predators should begin hatching out and the population will build up rapidly. After a week or so, a noticeable reduction in the amount of red spider mite should have occurred. If you cannot see this after two weeks, your introduction has probably failed and you should reorder.

The pest should be virtually wiped out by the control, which can stay alive for about three weeks without a food supply. If no new infestations occur within this time, the predator will die out and will need re-introducing to deal with any further attacks.

WHITEFLY

Like red spider mite, this pest attacks tomatoes, cucumbers and a variety of ornamental plants. It looks like a tiny white moth, and clouds of whiteflies arise when an infested plant is shaken lightly (Fig. 14). They feed by sap sucking, and excrete a sticky honeydew which coats the leaves and encourages the growth of a black fungus known as sooty mould.

Whiteflies have a more complex life cycle than red spider mite. Adults lay eggs which hatch into nymphs that look like scales. The nymphs move about the plant feeding, then they settle down and become fixed to the leaf while they go through several more stages of their life cycle before finally emerging as adults. While fixed to the leaf, the scale thickens and is impervious to chemicals, making whitefly very difficult to control by spraying, as only the adults are vulnerable.

The biological control for whitefly is a small wasp

Fig. 14 *The whitefly parasite turns whitefly scales black when they have been attacked.*

called *Encarsia formosa.* This is a parasite, not a predator, which means it is unlikely to completely wipe out its host. Adult females lay eggs inside the whitefly scales, one egg per scale. (Reproduction is parthenogenetic – it is not necessary for male wasps to fertilize the eggs.) The egg hatches out inside the scale, feeding on the larva within it: the scale turns black and the adult wasp eventually cuts a hole in the top of the scale to emerge.

Parasites should be ordered as soon as the first whitefly is seen. They are supplied not as live wasps, but as parasitized scales on pieces of tobacco leaf, usually stuck on cards. These should be hung on infested plants, evenly distributed, and out of direct sunlight as far as possible. The wasps are small and impossible to identify, but you can check that they have emerged from the parasitized scales by holding them up to the light and looking for the exit holes. After a couple of weeks, the first black scales should start appearing on your plants to show that the wasps are doing their job. Keep the greenhouse above 21°C (70°F) for the wasps to work most efficiently; the minimum to which it should be allowed to fall is 15°C (59°F). Below this level, the wasps reproduce and develop much more slowly.

CATERPILLARS

There are many types of caterpillar that damage plants, both in the garden and in the greenhouse. One of the best known is the cabbage white, the larva of butterfly, that can reduce a crop of cabbages or related plants to ribbons. The black and yellow caterpillars are too well known to need a description. In greenhouses, tomato moth can be a serious pest, eating the fruit and rendering it unusable.

The biological control for caterpillars is a bacterium, *Bacillus thuringiensis*. The bacterium paralyzes the gut of the caterpillar, preventing it feeding; it also releases toxins which will kill it. It is specific to butterfly and moth caterpillars (*Lepidoptera*) and will not harm other creatures. However, it must not be used indiscriminately as it will also kill the caterpillars of non-harmful butterfly and moth species. This is unlikely to happen as the bacterium must be applied to the food plant of the caterpillar, and you are not likely to spray nettles or other butterfly food plants in mistake for cabbages. The caterpillar needs to eat the bacterium to be affected – it would not be harmed by spray drift falling on it, for example.

Bacillus thuringiensis is supplied as a powder containing spores and toxins. It is mixed with water and applied to the target plants as a spray, ensuring both surfaces of the leaf are covered. While it will stop the caterpillars eating immediately they ingest it, they do not die for a few days, so don't be misled into thinking it hasn't worked.

Because the powder is sold in sealed sachets, it stores well and you can have it on hand for use as necessary – unlike other, live controls that have to be ordered as they are needed. It is effective on all *Lepidoptera*, but it will not kill the larvae of sawflies, although they look like caterpillars, because they are not in this group.

APHIDS

Biological controls for aphids are fairly new. There are two controls – a predator and a parasite – both intended for use in the greenhouse.

Aphids are very common pests, usually known as blackfly or greenfly. They can occur in vast numbers, sucking sap and weakening plants, distorting shoot tips and flower buds, making stems and foliage unpleasantly sticky with honeydew and probably most significant of all, spreading virus diseases.

Aphidoletes is a predatory midge which eats most types of aphid. It is available in spring and summer, and is supplied as small orange larvae. Open the package in the greenhouse and remove the larvae carefully with a paintbrush, placing them among the aphid colonies. Alternatively, you can just lay the plastic tube beneath an infested plant and wait for the larvae to crawl out. They feed on the aphids for several days, then fall to the ground to pupate in the soil. After two weeks, adults hatch out and continue the cycle. Check the orange larvae or tiny red eggs among the aphids to ensure the midges are breeding.

Aphidius is a parasitic wasp available all year round. It is a little more selective about the type of aphid it prefers, but is fast acting. Supplied as pupae ready to hatch, the package should again be opened in the greenhouse in case any wasps have hatched on the journey, and the open tube laid beneath an infested plant. Leave it there for a week to allow all the pupae to hatch. Within a short while, you should be able to see brown, mummified, parasitized aphids to show the wasps are at work.

Another method of controlling aphids is by the fungus *Verticillium lecanii*, which is not yet available to amateur gardeners. It needs warm, very humid conditions, and is introduced in the form of a spray before many aphids are present on the crop. Those aphids it kills become covered by white fluffy spores which will be released to infect further aphids.

MEALYBUGS

These pests look a little like woodlice covered with white fur. They are common on cacti, orchids, grapevines and several ornamental plants, both in the house and greenhouse. Because of their waxy, 'woolly' coating, they are shielded from insecticide sprays and are difficult to control with chemicals. They look unpleasant, weaken plants, and encourage the development of sooty moulds.

Control is by a predatory ladybird, *Cryptolaemus*, which originates from Australia. The black and brown adults are of typical ladybird appearance, but the larvae look very much like their prey. They are covered with a similar woolly coating, but they are a little larger. Both adults and larvae eat eggs and adult mealybugs.

The predators are supplied as a mixture of adults and larvae. As usual, open the package in the greenhouse so that the adults can fly straight out: the larvae should be handled very gently, lifting them carefully with a fine brush, or shaking them on to a clump of mealybugs. Keep the greenhouse between 20–25°C with high humidity levels.

The life cycle of the beetle takes several months to complete and they are quite expensive to rear. About 15 predators are recommended to establish a breeding colony in a greenhouse with a moderate infestation of mealybugs. The adults are not easy to spot and the larvae are very difficult to distinguish from their prey; a gradual reduction in pests is the only sign you will have that your introduction has been a success.

THRIPS

These pests are usually of fairly minor importance to gardeners, but are of more significance to commercial growers, particularly on sweet peppers and cucumbers. They are tiny insects known commonly as thunderflies, with fringed wings. They feed by sap sucking, causing mottling and flecking of foliage and sometimes damage to fruit and flowers. Western flower thrip has recently become of great commercial importance.

Amblyseuis cucumeris is a small mite rather similar to the red spider mite predator, and it feeds on thrips. It needs to be introduced immediately thrips are seen, and should keep the population low enough to prevent damage occurring. One introduction should be sufficient for the whole season.

LEAF MINER

There are several types of leaf miner which can cause a lot of damage to plant foliage. The adults are small flies, which lay eggs just below the leaf surface. When the larvae hatch, they eat the leaf tissue under the surface, causing characteristic mines or tunnels which disfigure the foliage and can weaken the plants. The larvae drop to the ground to pupate in the soil.

Two parasites, both ichneumon wasps, have recently been used to control leaf miner on commercial establishments. *Diglyphus isaea* kills the leaf miner larva in its mine and lays an egg beside the freshly killed food source. The larval wasp feeds in the mine, emerging as an adult. *Dacnsa sibirica* lays its egg inside the leaf miner larva itself; the larva continues to feed and drops to the soil to pupate, but the parasite then takes over and will emerge from the pupa instead of a leaf miner.

VINE WEEVILS

The main damage done by vine weevils is carried out by the larvae, which feed on the below-soil parts of a wide variety of ornamental pot plants. Cyclamen are particularly badly affected, and often the collapse of the plant is the first sign that anything is wrong – by which time, of course, it is too late to do anything about it. When a dying plant is turned out of its pot, half a dozen or more fat, white, C-shaped larvae can usually be seen in the compost. The adults can also cause damage to plants by biting characteristic notches from the leaf edges: rhododendrons are particularly susceptible. Adults are small and black, of typical weevil shape, and feed only at night.

Chemical control relied on the incorporation of long-lasting and often toxic chemicals into the compost as a preventive measure. Biological control uses a parasitic nematode, *Steinerma biblonis*, which occurs naturally in the soil. Nematodes are best known as plant pests, but this one is completely unrelated to any pest species. The nematodes search out vine weevil larvae in the compost and enter them through body openings.

Symbiotic bacteria released from the nematodes reproduce within the dead body and are then released into the compost to seek out more vine weevils.

The nematodes are supplied in polythene packs containing an inert carrier, and should be stored in a refrigerator if their use is delayed. The pack contents are suspended in water and watered on to the crop to be protected. They are best applied as a preventive measure, but they can be used as a treatment where vine weevil larvae are already present. They function best at temperatures of 15°C (50°F) or more. One application should protect plants for eight weeks.

SILVER LEAF

There are not too many biological controls for diseases (rather than pests) but there is one for silver leaf. This is a bacterial disease, mainly of plums, but it will affect all members of the family *Rosaceae*, including cherries, apples, almonds etc. The symptoms are a silvering of the leaves which can be rather difficult to see, general unthriftiness of the tree and a brown stain in the centre of the wood, which can be spotted easily when a branch is cut.

The control for silver leaf, at present available only to commercial growers, is a fungus called *Trichoderma*. It is supplied as impregnated wooden dowels or pellets which are inserted in the trunk of the tree in predrilled holes. These should be spaced every 10 cm (4 in) or so in a spiral round the trunk: most trees need 10–20 pellets.

Trichoderma is also available as a powder to be mixed with water to form a paste or spray: this is then applied to pruning cuts on all types of fruit trees to prevent the entry of various diseases. It is possible that *Trichoderma* will receive clearance for use by amateur gardeners in the near future.

THE FUTURE

New biological control agents are being sought all the time. A type of anthocorid bug called *Orius* is a universal predator which should be available shortly, and even weeds may be controlled by natural agents soon. Japanese knotweed, an extremely fast-growing and vigorous weed, is kept under control by natural agents in Japan and scientists are looking at ways to import those controls here.

The ideal biological control is specific to a pest species; easy to breed and distribute; does not require very specialized conditions; multiplies at the same rate or slightly faster than its prey; and is efficient at seeking out and controlling that prey. Organisms capable of fulfilling these conditions are continually being discovered.

CHAPTER 6

CHEMICALS IN THE GARDEN

Organic gardening relies mainly on preventing problems with pests and diseases – once they have occurred it is sometimes difficult to treat them without resorting to chemicals.

There are chemicals that are approved by organic gardening associations, however, because they are derived from plants and will break down naturally in the environment. Some of these can be bought; others could be made up at home. However, in some countries, the use of home-made remedies may be restricted by law.

ORGANIC PESTICIDES

The insecticides described here are all contact insecticides – in other words they must be sprayed directly on to the pests to have any effect.

DERRIS

This is made from the roots of a plant, and its proper chemical name is rotenone. It is sold both as a liquid and a dust. It breaks down quickly in the environment and is harmless to mammals, bees and hoverflies. However, it *is* harmful to fish, amphibians, adult ladybirds, anthocorid bugs, lacewings and some other beneficial creatures.

It can be used against aphids, small caterpillars, sawfly larvae, flea beetles and raspberry beetles. It is only moderately effective.

PYRETHRUM

Made from the flowers of a species of chrysanthemum, pyrethrum is available alone, or mixed with derris, as a liquid or a dust. It is often mixed with another chemical called piperonyl butoxide which increases its efficiency but makes it unacceptable to organic gardeners. Pyrethrum is harmless to mammals but damaging to fish, amphibians and many beneficial insects.

Use it against aphids, flea beetles and small caterpillars. It is rather more effective than derris; mixing derris and pyrethrum appears to make both chemicals work more effectively.

INSECTICIDAL SOAP

A soap made from naturally occurring fatty acids of vegetable origin. Traditional 'soft soap' has a similar action but is rather milder; both are harmless to mammals and some beneficial insects but damaging to other beneficial insects such as ladybirds.

Soap can be used against aphids, whitefly, red spider mite, mealy bug and scale insects, but it can damage some plants. Do not use it on seedlings, newly transplanted plants, sweet peas or begonias, and make a small test spraying before using it on a wide scale.

QUASSIA

The bark of a tropical tree provides chips which are boiled in water to produce a very bitter extract. It is sprayed on to aphids, small caterpillars and sawflies: it is also painted on the trunks of saplings to deter rabbits, mice and deer which would otherwise chew the bark. It is approved for use as an insecticide.

Quassia is harmless to mammals, ladybirds, bees and many other beneficial insects. It is moderately effective.

NICOTINE

No longer available in most countries because it is so highy toxic. Derived from tobacco plants, nicotine is very poisonous to mammals, fish, amphibians and bees, but not to ladybirds, lacewings and some other beneficial insects.

It was effective at killing aphids, caterpillars and several other pests. Nicotine extract has in the past been obtained by the rather disgusting operation of boiling up old cigarette ends.

GARLIC SPRAY

Cloves of garlic are crushed and soaked in liquid paraffin overnight. The strained oil is then mixed with soft soap and diluted with water before being sprayed on plants against a variety of ailments including peach leaf curl and blackspot. It is harmless to mammals and insects, and is not proven to do any harm to the diseases either.

ELDER LEAF SPRAY

Elder leaves are boiled in water and the extract mixed with soft soap and water. Apparently harmless to mammals and beneficial insects (though the elder tree does contain toxins). Used as a spray against fungal diseases, especially mildew. Of doubtful efficacy.

RHUBARB SPRAY

Made like elder leaf spray, rhubarb spray was used against a variety of diseases, also some pest species.

SULPHUR

A fungicide for use against a range of diseases such as mildew. Sulphur can harm some beneficial insects and some fruit varieties, so should be used with care; it is harmless to mammals.

BORDEAUX MIXTURE

A mixture of copper sulphate and lime for use against mildew, potato blight and blackspot. Harmless to mammals and most beneficial insects.

NON-ORGANIC – BUT SAFER

Two pesticides cannot be classed as organic, but because they are specific to their action and do not harm beneficial species you may consider using them.

ALUMINIUM SULPHATE

This powder acts on slugs and snails by shrinking the slime-producing organs so they cannot make mucus. It can be sprinkled on the soil or made into a liquid; it can sometimes damage plants. It is harmless to mammals and other beneficial creatures.

PIRIMICARB

This is a synthetic chemical which is a specific aphicide and does no harm to any beneficial creatures. It is a contact insecticide, but it also passes from the upper to lower surface of the leaf to some extent.

RULES FOR USING PESTICIDES

Store chemicals safely, in their original containers, away from children.